AF702997

PHARMACOLOGY - RESEARCH, SAFETY TESTING AND REGULATION

CORTICOSTEROIDS AND STEROID THERAPY

NEW RESEARCH

PHARMACOLOGY - RESEARCH, SAFETY TESTING AND REGULATION

Additional books in this series can be found on Nova's website under the Series tab.

Additional e-books in this series can be found on Nova's website under the e-book tab.

PHARMACOLOGY - RESEARCH, SAFETY TESTING AND REGULATION

CORTICOSTEROIDS AND STEROID THERAPY

NEW RESEARCH

CARMEN ADKINS
EDITOR

New York

Copyright © 2015 by Nova Science Publishers, Inc.

All rights reserved. No part of this book may be reproduced, stored in a retrieval system or transmitted in any form or by any means: electronic, electrostatic, magnetic, tape, mechanical photocopying, recording or otherwise without the written permission of the Publisher.

For permission to use material from this book please contact us:
nova.main@novapublishers.com

NOTICE TO THE READER

The Publisher has taken reasonable care in the preparation of this book, but makes no expressed or implied warranty of any kind and assumes no responsibility for any errors or omissions. No liability is assumed for incidental or consequential damages in connection with or arising out of information contained in this book. The Publisher shall not be liable for any special, consequential, or exemplary damages resulting, in whole or in part, from the readers' use of, or reliance upon, this material. Any parts of this book based on government reports are so indicated and copyright is claimed for those parts to the extent applicable to compilations of such works.

Independent verification should be sought for any data, advice or recommendations contained in this book. In addition, no responsibility is assumed by the publisher for any injury and/or damage to persons or property arising from any methods, products, instructions, ideas or otherwise contained in this publication.

This publication is designed to provide accurate and authoritative information with regard to the subject matter covered herein. It is sold with the clear understanding that the Publisher is not engaged in rendering legal or any other professional services. If legal or any other expert assistance is required, the services of a competent person should be sought. FROM A DECLARATION OF PARTICIPANTS JOINTLY ADOPTED BY A COMMITTEE OF THE AMERICAN BAR ASSOCIATION AND A COMMITTEE OF PUBLISHERS.

Additional color graphics may be available in the e-book version of this book.

Library of Congress Cataloging-in-Publication Data

ISBN: 978-1-63482-308-1
Library of Congress Control Number: 2015932939

Published by Nova Science Publishers, Inc. † New York

Contents

Preface

Corticosteroids (CS) are naturally occurring biomolecules produced in the adrenal cortex and have a multitude of roles which includes carbohydrate, protein and fat metabolism, inflammation and regulation of water, electrolyte etc. Based on their functions, steroids are classified as glucocorticoids and/or mineralocorticoids, and only the former have anti-inflammatory properties which have been chemically modified to produce potent anti-inflammatory drugs which also retain the metabolic and bone effects of the primary chemical. This book provides new research which includes the role of corticosteroids in diseases such as chronic obstructive pulmonary disease, adenoids, chronic subdural hematoma, osteonecrosis, and autoimmune diseases.

Chapter 1 – Chronic obstructive pulmonary disease (COPD) is mainly caused by smoking and presents with shortness of breath that is progressive and irreversible. In the third world use of biomass fuel has also been associated with COPD. It is a worldwide health problem and fourth most common cause of chronic disability and mortality even in developed countries. It is a complex disease in which both airway and lung parenchyma is involved. Inhaled corticosteroids (ICS) are widely used in clinical practice for the management of COPD however, their efficacy is still debated. They have shown beneficial effects on airway inflammation and infections and have also improved lung function and quality of life of COPD patients. There is epidemiological evidence that steroids might also protect against lung cancer in mild-moderate COPD but not so much in severe disease. This might be due to their effects on the process of epithelial mesenchymal transition (EMT), which is active in smokers and COPD. This opens up a new therapeutic area for the management/treatment of lung cancer in COPD. In this chapter the authors have reviewed the current literature on role of ICS in COPD; especially focusing on the effects of ICS on airway inflammation, infections, remodeling changes including matrix changes and EMT. The authors also reviewed the literature on effects of ICS on lung cancer risk in COPD.

Chapter 2 – Adenoid hypertrophy (AH) is a common childhood disease as associated with nasal obstruction with snoring, mouth breathing, hyponasal speech, rhinorrhea, and occasional abnormal facial development known as adenoid facies. By obstructing the rhinopharynx and nasopharyngeal orifice of the Eustachian tube, AH may also cause sinusitis and otitis media with effusion. Moreover, in the most serious cases, when associated with tonsillar hypertrophy, can cause obstructive sleep apnea syndrome. Treatment of AH in pediatric patients children depends on the degree of airway obstruction and related morbidity. Adenoidectomy has been traditionally considered to be definitive treatment for relief of upper airway obstruction and disease complicated by or attributable to AH. However, adenoidectomy has several pitfalls such as regrowth of adenoid after surgical removal, general complications (i.e., adverse anesthetic events and respiratory complications), and postoperative bleeding. A medical alternative to adenoidectomy is systemic steroid therapy, which leads to a prompt, temporary decrease in adenoid size, although chronic systemic administration is associated with serious adverse events.

An alternative to systemic steroids is the use of intranasal corticosteroids (INCS), which include beclomethasone dipropionate, budesonide, flunisolide, fluticasone propionate, and mometasone fluroate. INCS are associated with minimal systemic effects and they also have a substantially improved therapeutic index compared with intravenous and oral corticosteroids. The purpose of this review is to analyze the efficacy of each intranasal steroid and describe the spectrum of complications associated with their use.

Chapter 3 – Chronic subdural hematoma (CSDH) is a common condition in the elderly population and one of the most frequent lesions encountered in neurosurgical departments.

Mild head trauma is reported in most cases, but the pathophysiology of CSDH is still a matter of debate. Several data support the role of inflammatory related factors in the pathogenesis of the lesion, thus CSDH is considered a chronic self-perpetuating inflammatory process involving the dura matter.

Surgical treatment is the most common procedure for this type of lesion and it has proved to be effective. However, there is a large amount of data supporting the use of steroids in the management of CSDH. This data is essentially based on the inflammatory processes that have been postulated as underlying CSDH development.

The aim of this chapter is to describe the current role of steroids in the management of CSDH based on the pathophysiological processes that have been postulated as underlying CSDH development.

Chapter 4 – Osteonecrosis of femoral head (ONF) is one of the serious adverse events in the patients with systemic lupus erythematosus (SLE) associated with corticosteroid therapy. The authors have reported a multicenter prospective study of prevention of ONF in SLE patients on high doses of corticosteroids using anticoagulant of warfarin. In the diagnosis of ONF, plain radiography and magnetic resonance imaging (MRI) are important. Especially, in early stage of ONF, although the plain radiograph is still normal, evident changes can be seen in MRI. The treatment of ONF remains controversial. Anticoagulants may be useful to prevent ONF. Therefore, early diagnosis and prevention of ONF are critical issues especially in SLE patients. In this chapter, we present the radiological images illustrating osteonecrosis in patients with autoimmune diseases including SLE, and review the strategy to prevent ONF induced by corticosteroids.

Chapter 5 – Corticosteroids may sometimes exert unfavorable effects, thrombosis, avascular necrosis, endothelial cell damage, and corticosteroid vasculitis on the blood vessels in systemic autoimmune diseases. The association of corticosteroid and the thrombotic events has been reported. However, the mechanism of thrombotic tendency induced by corticosteroids has not been fully elucidated. Soluble endothelial cell protein C receptor (sEPCR) is one of the factors to regulate coagulation system. The authors found that sEPCR is a sensitive biomarker of endothelial injuries caused by active disease and often by corticosteroids in systemic lupus erythematosus (SLE). The findings of frequent events especially in the patients with SLE may illustrate the relationship between sEPCR and corticosteroids. sEPCR could be used as a predictable marker of vascular complications and thrombosis induced by corticosteroids in SLE.

In: Corticosteroids and Steroid Therapy
Editor: Carmen Adkins
ISBN: 978-1-63482-308-1
© 2015 Nova Science Publishers, Inc.

Chapter 1

Role of Corticosteroids in Chronic Obstructive Pulmonary Disease (COPD)

Mathew Suji Eapen*[1]*, Shakti Dhar Shukla*[1]*,
Malik Quasir Mahmood*[1]*,
Kielan McAlinden-Volkovickas*[1]*,
Rajaraman D. Eri*[2]*, Eugene Haydn Walters*[1]
and Sukhwinder Singh Sohal*[1 2,*]

[1] Breathe Well Centre of Research Excellence for Chronic Respiratory Disease and Lung Ageing, School of Medicine, University of Tasmania, Hobart, Tasmania, Australia

[2] School of Health Sciences, Faculty of Health, University of Tasmania, Launceston, Tasmania, Australia

Abstract

Chronic obstructive pulmonary disease (COPD) is mainly caused by smoking and presents with shortness of breath that is progressive and irreversible. In the third world use of biomass fuel has also been

* Corresponding Author: Dr Sukhwinder Singh Sohal; School of Health Sciences, Faculty of Health; University of Tasmania; Locked Bag 1322, Newnham; Launceston, Tasmania 7248; Australia; Email: sssohal@utas.edu.au.

associated with COPD. It is a worldwide health problem and fourth most common cause of chronic disability and mortality even in developed countries. It is a complex disease in which both airway and lung parenchyma is involved. Inhaled corticosteroids (ICS) are widely used in clinical practice for the management of COPD however, their efficacy is still debated. They have shown beneficial effects on airway inflammation & infections and have also improved lung function and quality of life of COPD patients. There is epidemiological evidence that steroids might also protect against lung cancer in mild-moderate COPD but not so much in severe disease. This might be due to their effects on the process of epithelial mesenchymal transition (EMT), which is active in smokers and COPD. This opens up a new therapeutic area for the management/ treatment of lung cancer in COPD. In this chapter we have reviewed the current literature on role of ICS in COPD; especially focusing on the effects of ICS on airway inflammation, infections, remodeling changes including matrix changes and EMT. We also reviewed the literature on effects of ICS on lung cancer risk in COPD.

Corticosteroids

Corticosteroids (CS) colloquially steroids are naturally occurring biomolecules produced in the adrenal cortex and have a multitude of roles which includes: carbohydrate, protein and fat metabolism, inflammation and regulation of water, electrolyte etc. Based on their functions steroids are classified as glucocorticoids and/or mineralocorticoids, only the former have anti-inflammatory properties which have been chemically modified to produce potent anti-inflammatory drugs which also retain the metabolic and bone effects of the primary chemical.

History

The earliest understanding of glucocorticoid (GC) as a form of medicine was based on the findings by Polish-Swiss scientist Tadeus Reichstein and two American scientists Philip S. Hench (1896-1965) and Edward C. Kendall (1886-1972). For their contribution they jointly shared the Nobel Prize award in medicine in 1950 (Shampo & Kyle; Shampo & Kyle; Shampo, Kyle, & Steensma). Reichstein and his colleagues had isolated and crystallised nine substances from the adrenal gland one of which was named *corticosterone* which was used to treat Addison's disease (Shampo & Kyle). Kendal discovered 28 different forms of cortical hormones from the adrenal glands and found that six of them were active and named these effective compounds as A, B, C, D, E and F. Compound A was synthesized in 1944 and compound E in 1946 which was later called cortisone (Shampo & Kyle). Hench in the meantime was researching a cure for rheumatoid arthritis. He found rheumatoid arthritis patients who additionally suffered from jaundice showed a lapse in the symptoms associated with cortisone and he factored that the unknown substance which he named as substance X was present in the blood and was responsible for this effect. Later on he found that a secretory factor from the adrenal gland was responsible. He teamed up with Kendal who provided him with compound E (cortisone) in sufficient amounts to be used to treat the patients with Rheumatoid Arthritis. Their research further lead to the discovery of ACTH (adrenocorticotropic hormone), a hormone of the pituitary gland, also waned of alleviation of symptoms of rheumatoid arthritis. Glucocorticoids have been considered as one of the most effective group of medication of the 20th century and have treated a wide range of inflammatory and allergic diseases which include various forms of arthritis, asthma (Spahn & Leung, 1996), chronic obstructive pulmonary disease (COPD) (Adcock & Ito, 2005), multiple sclerosis (Goodin, 2014), ulcerative colitis (Triadafilopoulos, 2014), and more.

Potency of Corticosteroids

Synthetic corticosteroids have been designed by modifying the 21 carbon steroid skeleton. Based on their structural modification the degree of potency, selectivity, duration of activity and protein binding can be altered. For example, when an additional bond between carbon-1 and carbon-2 is added in cortisol (hydrocortisone), both the glucocorticoid and anti-inflammatory activity is enhanced. Changes such as fluorination of the C-9 position enhance both the glucocorticoid as well as mineral-corticoid activity of the corticosteroids. Glucocorticoids are absorbed easily through human intestine and are thus generally administered orally but they have also been used in inhaled and topical formulations.

Mechanism of Action of Corticosteroids

CS can work in two different ways; at higher concentration they are associated with the activation of anti-inflammatory genes and at low doses they are associated with gene suppression by recruiting histone deacetylase (HDACs) to the sites of pro-inflammatory transcription (P. J. Barnes, 2006b).

There are 11 HDAC isoenzymes that deacetylase histones within the nucleus and specific HDACs appear to be differentially regulated to control different sets of genes. They are divided into two major classes. Class I comprises HDAC1, 2, 3, 8 and 11 whereas class II includes HDAC4, 5, 6, 7, 9 and 10. Marked reduction in HDACs has been observed in a variety of chronic inflammatory diseases (P. J. Barnes, Adcock, & Ito, 2005; de Ruijter, van Gennip, Caron, Kemp, & van Kuilenburg, 2003; Ficner, 2009), hence could be relatively CS un-responsive due to this.

In COPD a marked reduction in HDAC2 expression and activity has been observed, with rather less reduction in HDAC5 and HDAC8 expression, and normal expression of other HDACs. It appears that for the regulation of inflammatory genes HDAC2 appears to be of critical importance (P. J. Barnes, 2009; Ito et al., 2005). We recently reported that HDAC2 expression is increased in physiologically normal smokers (may be an anti-inflammatory response) but reduced in current smokers with COPD, though the latter finding is partly confounded by general decrease in cellularity in the lamina propria. Quitting smoking may well have a real effect on up-regulating HDAC2 at a cell level, as seen in COPD ex-smokers, but was not affected by ICS (Inhaled Corticosteroids) therapy. Interestingly no change was observed for HDAC2 expression in the epithelium in COPD current and ex-smokers, normal lung function smokers as compared to normal never smokers. Indeed, the main message may be that HDAC2 expression in the epithelium was pretty well preserved generally in smokers, suggesting that there is sufficient HDAC2 expression here to allow ICS to be effective in this compartment.

Molecular methods, as exclusively published in the past on this topic of HDAC2 in the airways, cannot fully take account of such changes in the cellular environment (Glare, Divjak, Bailey, & Walters, 2002) and cannot be interpreted on simple face value. However, the proposition that HDAC2 activity is decreased in the airway wall generally in COPD does seem to be correct, with major implications for understanding the aetiology of this common disease.

Gene Activation and Suppression by Corticosteroids

At high dose the activation of anti-inflammatory genes is mainly through binding to glucocorticoid receptors (GRs) localised in the cytoplasm of the target cell; the complex acts as a transcription factor to control the transcription of several steroid responsive genes (P. J. Barnes, 2006b). In the cytoplasm of the cells GRs are usually attached to proteins recognized as molecular chaperons, which include heat shock-proteins-90 (hsp-90) and FK binding protein. These proteins, when bound to the GC-receptor prevent its nuclear localisation by covering the sites of the receptor that are essential for transportation into the nucleus (Wu et al., 2004).

Binding of corticosteroids to GR leads to changes in the receptor configuration which further leads to dissociation of these inhibitory molecular chaperons proteins, exposing the sites essential for nuclear localisation. This results in rapid transport of active GR-complex into the nucleus, where it binds to glucocorticoid response elements (GRE) in the promoter region of steroid responsive genes, mainly types of anti-inflammatory genes leading to increase in synthesis of anti-inflammatory proteins such as annexin-1 (lipocortin-1), secretory leukocyte protease inhibitor (SLPI), IL-10, the inhibitor of NF-κB (IκB-α) glucocorticoid-induced leucine zipper protein, (which inhibits both NF-κB and AP-1) and mitogen-activated protein (MAP) kinase phosphatase-1, (which inhibits p38 MAP kinase) (P. J. Barnes, 2006b; P. J. Barnes et al., 2005). Activation of anti-inflammatory genes by high dose of corticosteroids is associated with a selective acetylation of lysine residues 5 and 16 on H4 (histone 4), resulting in increased gene transcription, whereas in response to inflammatory stimuli differential acetylation of residues 8 and 12 is involved. The fact that high doses of corticosteroids are needed for these actions put a major question mark against their clinical relevance. In clinical practice, corticosteroids are able to suppress inflammation at low doses (P. J. Barnes, 1998, 2006b, 2006d, 2009; P. J. Barnes et al., 2005).

Corticosteroid at lower doses leads to suppression of inflammatory genes by recruiting HDAC2 to activated pro-inflammatory transcriptional complexes. Initially it was believed that gene suppression by corticosteroids is generally induced by binding of GR to negative GRE sites in their promoter region, but later it was confirmed that this process is applicable to only a small number of genes, and this mechanism does not include genes encoding most inflammatory proteins (Ismaili & Garabedian, 2004). It was observed in asthma that most of the genes which are activated during the inflammatory process do not have GRE sites, but are still effectively repressed by corticosteroids (P. J. Barnes et al., 2005). There is convincing evidence now that anti-inflammatory actions of corticosteroids are due to inhibition of transcription factors such as AP-1 and NF-κB (by inhibition of histone acetylation and stimulation of histone deacetylation), which are transcription factors actively involved in regulation of many genes coding for a host of pro-inflammatory proteins (P. J. Barnes, 2006d; P. J. Barnes & Karin, 1997).

As already described, corticosteroids at low concentrations activate GRs which rapidly move to the nucleus and bind to co-activators such as CBP or PCAF to directly inhibit intrinsic HAT (histone acetyl transferase) activity, so that HDACs are recruited leading to histone deacetylation and suppression of inflammatory genes. So in very general terms, corticosteroids at low concentrations recruit HDACs to the transcription complex and convert the process of acetylation to deacetylation to suppress the transcription of inflammatory genes (P. J. Barnes, 2006a, 2006b, 2006c, 2006d, 2008, 2009; P. J. Barnes et al., 2005; P. J. Barnes & Karin, 1997).

Corticosteroids and Airway Inflammation in COPD

Chronic inflammatory diseases like COPD, asthma, cystic fibrosis, interstitial lung disease, inflammatory bowel disease and rheumatoid arthritis are associated with a specific pattern of inflammation, which requires a coordinate expression of a wide range of different pro-inflammatory genes, coding for various pro-inflammatory mediators.

In COPD, the specific pattern of inflammation is mainly characterized by increased numbers of luminal, sputum and BAL (bronchial alveolar lavage) neutrophils and airway wall macrophages and perhaps T-lymphocytes, predominantly cytotoxic (CD8+) cells (P. J. Barnes et al., 2005). Steroids do improve lung function, effect inflammation and structural changes in COPD. However, type of inflammation and structural changes effected by steroids are not well understood, and there are very few studies reporting the effects of ICS on airway inflammation and even fewer on structural changes (airway remodelling) in COPD. Overall our knowledge is limited (Chanez et al., 2004).

Hattotuwa and colleagues reported a significant reduction in CD4/CD8 cell ratios in the epithelium, and in subepithelial mast cells in bronchial biopsies obtained from COPD patients in the active ICS arm treated with fluticasone propionate compared to placebo (Hattotuwa, Gizycki, Ansari, Jeffery, & Barnes, 2002). Reid and et al. reported in both bronchial biopsies and BAL from COPD patients, showing fluticasone propionate reduced BAL neutrophils and BAL epithelial cell numbers, and CD68+ macrophages, CD8+ lymphocytes and mast cells in bronchial biopsies, but interestingly noted increased neutrophils in bronchial biopsies (Reid et al., 2008) suggesting sequestration rather than movement into the airway lumen. In another study, it was reported that three month treatment with fluticasone propionate significantly decreased mucosal mast cells and increased neutrophils, again in biopsies from COPD patients (Gizycki, Hattotuwa, Barnes, & Jeffery, 2002).

More recently, in a very interesting study, Lapperre et al. reported the effects of fluticasone propionate on inflammation in airway biopsies and sputum from COPD patients, and found that fluticasone significantly decreased the number of mucosal CD3+ cells, CD4+ cells, CD8+ cells and mast cells after 3 months, with effects maintained to 30 months. They also reported in airway biopsies that treatment with fluticasone for 30 months reduced the number of mast cells, increased number of eosinophil and increased the percentage of intact epithelium. This is the only study to our knowledge reporting effects of ICS on epithelium. There was also a decrease in sputum neutrophils, macrophages, and lymphocyte accompanied by improvements in FEV1 decline, dyspnea, and quality of life. The decrease in inflammatory cells correlated with clinical improvements. Discontinuing fluticasone for 6 months on the other hand increased CD3+ cells, mast cells, and plasma cells and thus was accompanied by deterioration in clinical outcomes (Lapperre et al., 2009). Thus, taking all of these intervention pathology studies together there is strong evidence for an anti-inflammatory cellular response to CS in COPD, with the exception being an effect of increasing neutrophils and perhaps (eosinophils).

Effects of Inhaled Corticosteroids on Airway Infections

One of the major risk factors for COPD is chronic respiratory infections, especially with bacteria such as *Streptococcus pneumoniae* and *Haemophilus influenzae* and viruses (mainly rhinoviruses), which are commonly detected in almost all the patients (Banerjee, Khair, & Honeybourne, 2004; Cabello et al., 1997; Chin et al., 2005; Sethi, 2004; Soler et al., 1998; Wilkinson et al., 2006).

The vast majority of COPD patients (~80%) use inhaled corticosteroids (Thomas, Radwan, Stonham, & Marshall, 2014), either alone or in combination with a long-acting β2 agonist (Decramer, Janssens, & Miravitlles, 2012). However, to date, the clinical studies on effect of inhaled corticosteroids (ICS) on airway infections have been small to detect important effect of these medications on infections in COPD and somewhat conflicting data has been reported (Lydia Finney et al.). More precisely, two studies using sputum cultures found no association between inhaled corticosteroid use and bacterial infection (Marin et al., 2012; Miravitlles et al., 2010). However, another study based on quantitative PCR (qPCR) reported that higher doses of inhaled corticosteroids were associated with greater bacterial loads (Garcha et al., 2012). This may be attribute to reduction in inflammatory profile as one study suggested that prolonged therapy with inhaled corticosteroids is effective in reducing the total sputum cell counts including epithelial cells, neutrophils and lymphocytes in stable COPD, only if higher cumulative dose (≥60 mg) or longer duration of therapy (≥6 weeks) was observed (Gan, Man, & Sin, 2005). However, there was no significant effect on sputum eosinophils and IL-8 and a very little effect on sputum macrophage population (Gan et al., 2005). Despite having only marginal reductions in sputum cell profiles, it may be useful in reducing hospitalizations due to exacerbations as well as reduction in sputum production and cough (Burge et al., 2000).

In bronchoalveolar lavage (BAL), ICS reduced neutrophil counts (SMD=−0.64 units, 95% CI: −1.05 to −0.22; P=0.003) and lymphocyte counts (SMD=−0.64 units, 95% CI: −1.13 to −0.15; P=0.01). Interestingly, ICS increased macrophage counts in BAL (SMD=0.68 units, 95% CI: 0.25 to 1.11; P= 0.002). The study did not find any clinically relevant effects on eosinophil counts (Gan et al., 2005).

In bronchial biopsies, ICS did not significantly affect neutrophil counts (SMD=0.61 units, 95% CI: −0.11 to 1.33; P=0.10). However, ICS reduced the CD8 lymphocyte counts (SMD=−0.66 units, 95% CI: −1.09 to −0.24; P=0.002) and the CD4 lymphocyte counts in the biopsies (SMD=−0.52 units, 95% CI: −0.79 to −0.25; P= 0.001). ICS did not have a significant effect on tissue CD68 macrophage counts (SMD=−0.32 units, 95% CI: −0.73 to 0.09; P=0.13). Again, the investigators reported no significant effects on eosinophil counts (Gan et al., 2005).

One major concern of prescribing ICS to COPD patients has been significantly higher rates of pneumonia associated with the use of ICS, as reported by several clinical trials (Burge et al., 2000; Calverley et al., 2007; Crim et al., 2009). Findings from the TORCH study showed significantly increased incidence of pneumonia in ICS groups, with the probability of pneumonia being 12·9% with placebo, 13·3% with long-acting β2-agonist (LABA) monotherapy, 18·3% with inhaled corticosteroid monotherapy, and 19·6% in the inhaled corticosteroid in combination with LABA group. Although ICS treatment was found to be associated with an increased relative risk (RR) of pneumonia of 1·52 (95% CI 1·32–1·76), there was no significant increase in pneumonia mortality (Crim et al., 2009). Similarly, the INSPIRE study compared fluticasone combined with salmeterol with the long-acting muscarinic antagonist tiotropium. They reported the probability of having pneumonia within 2 years as 9·4% in the inhaled corticosteroids plus long-acting β2-agonist group and 4·9% in the tiotropium group (Calverley et al., 2011). Moreover, several population-based COPD-cohort studies have shown an increased risk of pneumonia with ICS, with an estimated RR of between 11% and 70%. However, much smaller (127 patients) case-control study reported a RR 3·26 (95% CI 1·07–9·98) (Singh, Amin, & Loke, 2009).

One study reported impaired clearance of *Klebsiella pneumoniae* in mice causing increased mortality due to fluticasone (Patterson, Morrison, D'Souza, Teng, & Happel, 2012), and in a mouse model of allergic airway disease, budesonide impaired host defense to *Pseudomonas aeruginosa* (P. Wang et al., 2013). Conversely, in other animal models and in in-vitro cell culture models, fluticasone reduced cellular adherence of *S. pneumoniae*, *H. influenzae*, and *P. aeruginosa* (Barbier, Agusti, & Alberti, 2008; Dowling, Johnson, Cole, & Wilson, 1999). Surprisingly use of inhaled corticosteroid in children with asthma was associated with increased pharyngeal carriage of *S. pneumoniae* (L. Zhang et al., 2013). In an observational study on COPD patients, prior use of ICS is reported to be independently associated with decreased risk of short-term mortality and use of mechanical ventilation after hospitalization for pneumonia (Cheng, Su, Wang, Perng, & Yang, 2014).

Quite recently, novel molecular techniques have recognized a wide range of bacteria that are likely to update our understanding of the role of microorganisms in the pathogenesis of COPD. It has been reported that lower respiratory tract is colonized by a ‘microbiome’ even in healthy individuals. One study reported that COPD is not associated with an alteration of the respiratory microbiome, whereas others have reported changes in relative abundance of specific microbial phyla and in microbial diversity, and inhaled corticosteroid use alters the microbiome in patients with COPD (Cabrera-Rubio et al., 2012; Erb-Downward et al., 2011; Han et al., 2012; Pragman, Kim, Reilly, Wendt, & Isaacson, 2012; Sze et al., 2012). These findings might be relevant to the effects of inhaled corticosteroid on exacerbations, especially pneumonia, and future studies are likely to yield further insights into the effects of inhaled corticosteroids on the respiratory microbiome.

Effects of Inhaled Corticosteroids on Exacerbations in COPD

The major therapeutic aim in COPD is to prevent exacerbations. The use of ICS is widespread in patients with COPD due to its ability to reduce exacerbations. Recent studies have concluded a beneficial effect of ICS in reducing the number of COPD exacerbations for patients, especially with advanced disease (FEV_1 <50% predicted) (Calverley, 2004). Several clinical trials strongly indicate that the use of ICS may reduce clinically relevant exacerbations by approximately 30% and improve health status of patients who have moderate to severe disease (Jen, Rennard, & Sin, 2012). Moreover, the withdrawal of ICS may likely result in increased risk of exacerbations and worsening of health status (Price, Yawn, Brusselle, & Rossi, 2013).

One study has observed comparable exacerbation rates between extrafine beclomethasone and fluticasone cohorts during the 2-year follow-up period. Odds of treatment stability (no exacerbation and/or treatment change) were significantly greater for patients initiating extrafine beclomethasone compared with fluticasone (adjusted OR 2.50; 95% confidence interval: 1.32–4.73). Moreover, median ICS dose exposure during study period of 2 years was significantly lower ($p<0.001$) for extrafine beclomethasone than fluticasone cohorts (315 μg/day vs. 436 μg/day for initiation, 438 μg/day vs. 534 μg/day for step-up patients) (Ceylan, 2006).

A recent Cochrane review suggested marginal positive effect of LABA/ICS inhalers on exacerbation rates in COPD patients in comparison to those with LABA alone. Data was reviewed from nine different studies, which randomised 9921 participants (rate ratio 0.76; 95% CI: 0.68 to 0.84) and corresponds to one exacerbation per person per year on LABA and 0.76 exacerbations per person per year on ICS/ LABA. Moreover, fluticasone plus salmeterol (FPS) lowered the odds of an exacerbation (OR=0.83, 95% CI: 0.70 to 0.98, 6 studies, 3357 participants). They concluded the risk of an exacerbation of 47% in the LABA group over one year, whereas 42% of people treated with LABA/ICS would be expected to experience an episode of exacerbation.

In addition, there was no significant difference in the rate of hospitalizations (rate ratio 0.79; 95% CI: 0.55 to 1.13) and was considered of very low quality evidence due to risk of bias, statistical imprecision and inconsistency across the studies (Kew, Mavergames, & Walters, 2013).

Effects Corticosteroids on Lung Function in COPD

Effect with corticosteroids especially in COPD and in its inhaled form has been variable and conflicting. Yang et al., 2012 (I. A. Yang, Clarke, Sim, & Fong, 2012) Cochrane collaboration report conducted meta-analysis of 55 randomised clinical studies which involved monitoring 16,154 stable COPD patients showed that although there was a decrease in exacerbation rate, the lung function parameter forced expiratory volume in one second (FEV(1)) remained unchanged. Similarly, four large studies that monitored the impact of long-term inhaled corticosteroids on lung function outcomes in patients with COPD, the Copenhagen City Heart Study, European Respiratory Society Study on Chronic Obstructive Pulmonary Disease, Inhaled Steroids in Obstructive Lung Disease (ISOLDE)(Vestbo et al., 1999), and Lung Health Study II, also showed no significant change in lung function over the two to three years (2000; Pauwels et al., 1999).

Contrary reports from Celli et al. showed that subjects in the Towards a Revolution in COPD Health (TORCH) study with moderate-to-severe COPD (mean postbronchodilator FEV~ 45% of predicted value), treatment with fluticasone propionate for 3 years had a reduction in the rate of decrease in FEV1 compared with placebo group (42 vs 55 mL/y, $P < .003$) (Celli et al., 2008). Also in ISOLDE studies like the TORCH study with fluticasone showed a reduction in exacerbation especially in patients with low lung function however there were no significant effect in terms of survival benefits.

Overall with clinical studies with ICS treatments has shown that there is a reduction in the exacerbation but no effect on mortality. Meta-analysis studies have shown that though corticosteroid is an effective treatment for exacerbations, there are some major problems associated with increased infection rates in patients undergoing the treatment.

Effects of Inhaled Corticosteroids on Quality of Life (QoL)

COPD exacerbations have a major impact on well-being and daily life, which is often measured by St. George's Respiratory Questionnaire (SGRQ) that has been validated for use in COPD patients (Jones, Quirk, Baveystock, & Littlejohns, 1992). A reduction in score of 4 is considered a clinically relevant improvement, reported by the patient and conversely, a worse quality of life predicts a worse clinical outcome (Osman, Godden, Friend, Legge, & Douglas, 1997). Adding ICS to the pre-existing treatment has been shown to significantly improve QoL in patients with COPD including improvement in lung function and decrease in respiratory symptoms significantly (Yildiz, Basyigit, Yildirim, Boyaci, & Ilgazli, 2004). However, there is limited reversibility of impaired lung function in COPD (Ceylan, 2006).

Budesonide–formoterol (BDF, inhalation powder) combination therapy provided significant clinical improvements in mean SGRQ total score than in placebo recipients (–3.9 vs. –0.03 units). Quite encouraging improvements in the SGRQ symptom and impact scores were observed in COPD patients (-5.9 and -4.7 units) receiving BDF, than the reduction in the SGRQ symptom and impact scores in placebo recipients (Szafranski et al., 2003). Moreover BDF combination therapy provided a persistent reduction of –7.5 in SGRQ score, compared with placebo. In addition, BDF improved SGRQ scores significantly when compared with only budesonide or formoterol, by 4.5 and 3.4 units, respectively (Calverley et al., 2003).

ICS/LABA was found to be more effective than LABA alone in improving health- related quality of life measured by the SGRQ (1.58 units lower with FPS; 2.69 units lower with BDF), dyspnoea (0.09 units lower with FPS), symptoms (0.07 units lower with BDF), rescue medication (0.38 puffs per day fewer with FPS, 0.33 puffs per day fewer with BDF), and FEV_1 (70 mL higher with FPS, 50 mL higher with BDF). Also, candidiasis (OR 3.75) and upper respiratory tract infection (OR 1.32) occurred more frequently with FPS than salmeterol alone (Nannini, Poole, Milan, & Kesterton, 2013). Despite all the beneficial effects of ICS in COPD, novel studies are required to understand the effect ICS may have on respiratory infections.

Effects of Corticosteroids on Airway Remodelling in COPD

Airway remodelling is defined as an alteration in size, mass or number of structural component of the lung tissue that leads to thickening of the airway wall because of increase in volume of epithelium, lamina propria, muscular or adventitial compartments particularly of small airways, which may leads to fixed airflow limitation in COPD and occurs early in the disease [1]. Data on effects of ICS on remodelling changes in COPD are very sparse. However, a few studies have shown that ICS can have significant effects on airway structural changes in COPD [16]. In this section we will briefly review the literature regarding effect of steroids on various aspects of airway remodelling.

Epithelium

The epithelium of the airways is extensively exposed to inhaled oxidants and irritants in smoke and thus is hugely exacerbated in those who habitually smoke cigarettes. The classic epithelial changes are mainly thickening of epithelial cells, general loss of epithelial cilia and metaplasia, goblet cell hyperplasia and squamous metaplasia i.e., loss of the normal pseudostratified structure to a thickened and squamous structure.

The major basic histological change in epithelium is its activation, which might be the first step in initiation of the cascade of downstream pathways for the development of COPD. In this context increased expression of epidermal growth factor receptor (EGFR) a marker of epithelial activation has been reported in COPD particularly in those who continue to smoke, which shows that airways epithelial cells are constantly being stimulated and activated by an insult like smoke (S. S. Sohal et al., 2010). Sohal et al. recently reported that high dose inhaled fluticasone propionate decreased airway EGFR expression in both current and ex-smokers with COPD (S. S. Sohal et al., 2014).

Airway epithelium is part of local innate immunity. It is activated through TLR (Toll-like receptors) and TNF-α production in response to microorganisms, allergens and pollutants. A study by Ning et al. reported that glucocorticoid enhances local innate host defence responses by epithelial expression of complement and other antimicrobial proteins. This is partially mediated through activation of transcription factor C/EBP in epithelial cells (N. Zhang, Truong-Tran, Tancowny, Harris, & Schleimer, 2007). Ruth et al.

further added that epithelial TLR4 and HBD2 expression gradually reduce with increasing severity of COPD and this effect only can be abrogated by combined ICS and LABA therapy. On the other hand, ICS given alone can impair the host defence against microbes by down-regulation of the TLR4 receptor while LABA plays a vital role in cAMP mediated post translation nuclear localization of TLR4 in bronchial epithelium (MacRedmond, Greene, Dorscheid, McElvaney, & O'Neill, 2007). It has also been observed that combined LABA and ICS potentiate the suppression of cigarette smoke induced IL-8 production by macrophages as well (Sarir et al., 2007).

Contradicting the largely positive picture from these studies, Michael and colleague in a vitro study reported that during viral infections ICS/LABA supress virus specific T-cell migration to lung epithelium and promoted virus propagation (and enhanced the clinical illness rather than suppressing it) (Edwards, Johnson, & Johnston, 2006). On the other hand in a study by Kan-o et al, in virally stimulated airway cells, ICS and LABA combination attenuated the virus-associated airway epithelial B7-H1 (co stimulatory molecule implicated in an escape mechanism by virus) via suppression of NF-kB activation. This in turn reduces the chance of the virus provoking exacerbations in COPD (Kan et al., 2013)

Reticular Basement Membrane (Rbm)

The human airway epithelium is attached to and supported by true basement membrane. The reticular basement membrane, also known as lamina reticularis, is a condensation of the extracellular matrix material, located just below the true basement membrane (Postma & Timens, 2006; Saglani et al., 2006), and separates the epithelium from the underlying lamina propria mesenchymal structure.

The baseline Rbm thickness in COPD is in doubt, with some studies reporting abnormally thick Rbm and some not (Chanez et al., 2004; Liesker et al., 2009; Postma & Timens, 2006).Our observation would suggest that it is thickened but variably so unlike the uniform appearance in asthma (S. S. Sohal et al., 2010; A Soltani, HK Muller et al., 2012). Until our work (below), the effects of ICS on the Rbm have not been specifically studied in COPD. However, we have reported that the Rbm in large airways from smokers and COPD patients is highly fragmented with elongated clefts evident with cells within these clefts and also hyper-vascularity throughout the Rbm (S. S. Sohal et al., 2010; Soltani et al., 2010). In a recent study we observed that inhaled fluticasone propionate normalized the Rbm fragmentation, and also decreased the Rbm cellularity both in COPD current smokers and ex-smokers. However, there was no effect on the hyper-vascularity of the Rbm (Singh Sukhwinder Sohal, Mahmood, & Walters, 2014).

Extracellular Matrix

The cell type that is involved in the ECM (Extracellular matrix) production is fibroblasts. The cells are of mesenchymal origin and have spindle shaped morphology. The active and secretory form of the fibroblasts is the myofibroblasts. The ECM consists of three major components: collagens, proteoglycans and elastic fibres, which are involved in cell migration, proliferation, adhesion, water balance and regulation of inflammatory mediators. In COPD patients ECM is affected dramatically with increased levels of matrix biomolecules and an irregular pattern of deposition leading to stiffness and reduced air flow.

Clinical studies showing any effect of corticosteroid on ECM changes are few and evidence is mainly from *in vitro* cell fibroblast and airway smooth muscle cultures and *in vivo* animal models. Recently a randomised clinical trial conducted by Hiemstra PS, Postma D and collaborators reported an increase in versican and collagen III deposition in large airway bronchial biopsies from COPD patients who had undergone a thirty month ICS treatment regimen, there was also a positive co-relation between lung function and increased collagen I expression in these patients when compared to the placebo (Kunz et al., 2013). *In vitro* studies have looked at opposing effects of combination therapy of corticosteroids and LABA on modulating lung fibroblasts to produce collagen and glycoproteins such as fibronectin and tenacin C. Corticosteroid in an inflammatory (high serum) environment, stimulated fibroblasts to myofibroblast, increasing collagen deposition and increased mRNA expression of COL4A1 and CTGF. However, LABA such as salmeterol and formoterol on their own were found to down-regulate these events. In a non-inflammatory or reduced serum condition corticosteroid down-regulated collagen deposition, heat shock protein 47 (Hsp47), and Fli1 mRNA expression (Goulet, Bihl, Gambazzi, Tamm, & Roth, 2007). Thus corticosteroids could play an important role in regulation of fibroblasts. However neither LABA nor corticosteroid was found to have no effect on fibroblast produced proteoglycans such as fibronectin and tenacin C (Degen et al., 2009; Goulet et al., 2007; Vanacker, Palmans, Pauwels, & Kips, 2002)

Smooth Muscle

Bronchial smooth muscle mass and its contribution to airway wall remodelling have been predominantly studied in small airways compared to large airways in COPD (P. K. Jeffery, 2001). Whether airway smooth muscle mass is a prominent cause of airflow limitation is still a matter of debate, with no convincing evidence really available (Kim, Rogers, & Criner, 2008). The cause of small airway narrowing in COPD is likely to be due to fibrosis and obliteration (Hogg et al., 2004).

However, an increase in airway smooth muscle in small airways was inversely correlated with lung function in one study (Chung, 2005). In another it was reported that muscle mass was increased by 50% in patients with severe COPD in small airways (Hogg et al., 2004). There is little information available on airway smooth muscle cells, but they may be functionally altered in proximal airways in COPD (Chung, 2005). Unfortunately there is no information available on function of the airway smooth muscle in small airways. It is not clear whether any increase in muscle mass in COPD is due to an increase in number of muscle cells, or increase in airway smooth muscle size, or a combination of both. In asthma, airway smooth muscle predominantly increases in large airways whereas in COPD this may occur mainly in small airways (Chung, 2005).

Abnormalities associated with smooth muscle mass in large airways have not been reported in COPD, although internal airway wall thickness has been associated with reduced FEV1/FVC ratio (Chung, 2005; Peter K. Jeffery, 2004). Biopsy studies from large airways have reported no increase in smooth muscle area and size; moreover, smooth muscle protein isoforms were not increased, but there was a slight increase in myosin light chain kinase but with no increase in phosphorylated myosin light chain (Chung, 2005).

In asthma, a beneficial role of ICS on airway smooth muscle has been reported, where airway smooth muscle thickening is prevented both in vivo and in vitro by ICS (S. Y. Lee et al., 2008; Vlahos et al., 2003). At the same time promising role of ICS over Salmeterol has been suggested in reducing chemokine production from human airway smooth muscle cells in asthmatics (John et al., 2004). Oltmanns et al. reported that fluticasone, but not salmeterol, has the potential to reduce the cigarette smoke-induced production of interleukin-8 in human airway smooth muscle from COPD patients (Oltmanns et al., 2008). In COPD, reduced FEV_1/FVC is linked with increased airway wall thickness (Peter K. Jeffery, 2004) but a clear impact of reduced airway smooth muscle mass with use of ICS has not been established so far. This warrants further study in COPD.

Goblet Cells, Sub-Mucosal Glands and Mucus

Mucus hyper-secretion in both asthma and COPD are therapeutically controlled by the use of mucolytics, anti-cholinergics and β2-Agonists. There is also a role for corticosteroid in in reducing mucus hyper-secretions thought to be through their direct action on controlling inflammation by reducing the activity neutrophils, eosinophils and other granulocytes (Hattotuwa et al., 2002; Innes et al., 2009). In allergic asthma and rhinitis patients glucocorticoids have been shown be effective in attenuating mucin production in this way (Peter J. Barnes & Pedersen, 1993; Weiner, Abramson, & Puy, 1998). In COPD inhaled corticosteroids in several large clinical studies mentioned earlier, have shown to reduce exacerbation rate improving the quality of life, however a direct correlation with inhibiting mucin production has yet to be established. However recently *invitro* studies have shown that glucocorticoids can decrease the expression of mucin in bronchial epithelial cells by supressing MUC5AC gene expression (Chen, Nickola, DiFronzo, Colberg-Poley, & Rose, 2006). Related studies using budesonide and fluticasone further proved suppression of MUC5AC protein expression in bronchial epithelial cells when induced by a combination of TGF-α and polyI:C in a dose-dependent manner (Takami et al., 2012). These actions show the potential of glucocorticoids as a promising therapeutic candidate for inhibiting mucin hyper-secretion but further studies are required to validate their effects clinically.

Vascular Remodelling

Data on airway vascular changes in COPD are very limited and studies of effects of ICS are more so. Kranenburg and colleagues reported enhanced expression of VEGF (vascular endothelial growth factor) in the bronchial, bronchiolar and alveolar epithelium and macrophages. They also observed VEGF expression in airway smooth muscle and vascular smooth muscle cells in both the bronchiolar and alveolar compartments (Kranenburg, de Boer, Alagappan, Sterk, & Sharma, 2005). They suggested that VEGF and its receptors VEGFR1 (decoy) and VEGFR2 (active) may be involved in epithelial and endothelial cell repair and maintenance in response to injury caused by cigarette smoking and may be involved with airway remodelling in COPD (Kranenburg et al., 2005; Siafakas, Antoniou, & Tzortzaki, 2007). Calabrese and colleagues (Calabrese et al., 2006) evaluated the contribution of vascularity in airway remodelling in smokers with normal lung function and smokers with COPD. They performed an immunohistochemical analysis involving vessels positive for integrin αvβ3. High αvβ3 expression was observed in bronchial vessels which was associated with higher cellular expression of VEGF, suggesting that these two molecules might be playing an important interacting role in angiogenesis (Calabrese et al., 2006).

In large airway biopsies, we recently reported that airway Rbm in current smokers with and without COPD is hyper-vascular; the lamina propria was hypo-vascular. COPD ex-smokers were close to normal in this regard, that vascular changes are secondary to smoking itself rather than presence of COPD (A Soltani, HK Muller et al., 2012; Soltani et al., 2010). In the Rbm TGF-β1 expression was increased in the abnormal vessels, which are also hyper-permeable (A Soltani, SS Sohal et al., 2012). Differential staining with factor VIII and collagen IV suggest that vessels are relatively new in the Rbm and old in the LP (Lamina propria) (A. Soltani et al., 2012).

However, we could not find any effects of inhaled fluticasone propionate on Rbm vessels in both COPD current or ex-smokers. However, in the LP fluticasone restored the vascularity to normal after six months of treatment in COPD current smokers (Soltani, 2010 #769; Soltani A, 2013 #770).

In a cross-sectional analysis, Zanini et al. reported that COPD patients treated with nebulised beclomethasone dipropionate show decreased bronchial vascular area and vessel size, and in VEGF, bFGF and TGF-β1 positive cells compared to the untreated. However their study did not show any abnormal change in the number of vessels in the lamina propria in smokers (Zanini, 2009 #394; Chetta, 2012 #771). More work is obviously needed to resolve these contradictions.

Epithelial Mesenchymal Transition (EMT)

EMT is a biological process in which epithelial cells undergo extensive molecular reprogramming and biochemical changes to acquire a mesenchymal phenotype. This is accompanied by progressive loss of epithelial markers such as cytokeratin(s), E-cadherins and simultaneous gain of mesenchymal markers such as S100A4, vimentin and N-cadherins, with increased migratory potential and invasiveness, and enhanced capacity to produce extracellular matrix components. EMT is a vital process during embryogenesis (Type I EMT), but can also be induced as a result of persistent insult and tissue inflammation (R. Kalluri, 2009; R. Kalluri & Weinberg, 2009; Zeisberg & Neilson, 2009). There are then two subsequent outcome possibilities with active EMT: severe and even complete organ fibrosis (Type II EMT), or development of a pre-malignant stroma when associated with angiogenesis (Type III EMT) (R. Kalluri, 2009; R. Kalluri & Weinberg, 2009; S. S. Sohal et al., 2013; S. S. Sohal & Walters, 2013a, 2013b; S. S. Sohal et al., 2013; S. S. Sohal, Ward, & Walters, 2014; Soltani et al., 2010; A Soltani, SS Sohal et al., 2012; Zeisberg & Neilson, 2009).

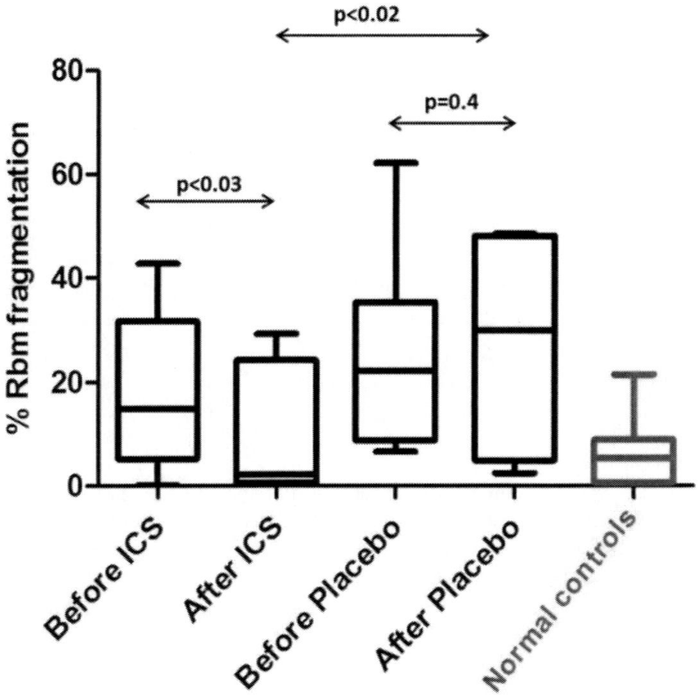

Figure 1. Percentage Reticular basement membrane (Rbm) fragmentation. Rbm fragmentation as % of total Rbm length before and after ICS versus before and after placebo, with normal control data for comparison. Post-treatment, the active treatment arm had significantly less splitting than the placebo group ($p<0.02$). After treatment fragmentation was normalized compared to normal controls ($p=0.4$). Data are represented as medians and ranges (S. S. Sohal et al., 2014).

We recently published that EMT is active in large airways of COPD patients (S. S. Sohal et al., 2010; S. S. Sohal et al., 2011). Furthermore, the Rbm in large airways is hyper-vascular (Amir Soltani, 2012; Soltani et al., 2010; A Soltani, SS Sohal et al., 2012) i.e., give the appearance of active EMT type-III. Cancer formation is common in COPD, and in these large airways especially squamous cell carcinoma (Raghu Kalluri & Neilson, 2003; R. Kalluri & Weinberg, 2009; S. S. Sohal, 2014; S. S. Sohal, A. Soltani et al., 2013; L. Yang et al., 2014). There is no sign of hyper-vascularity of the small airway Rbm in COPD, i.e., typical of EMT type-II (S. S. Sohal & Walters, 2013b); again it is small airways where the classic fibrotic changes occur.

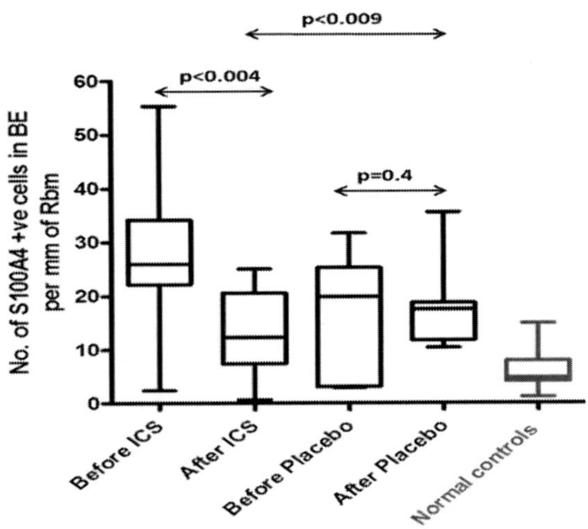

Figure 2. S100A4 in basal epithelium (BE). Number of S100A4 positive cells in the basal epithelium (BE) before and after ICS versus before and after placebo, with normal control data for comparison. Changes over time with ICS were also significant compared to those on placebo ($p<0.009$). After treatment, the active group was significantly different to normal controls ($p<0.02$). Data are represented as medians and ranges (S. S. Sohal et al., 2014).

There aren't any studies reporting effects of ICS on EMT in COPD. We were the first to undertake a study of ICS on EMT in the airways. In a randomized controlled trial we have recently reported that the inhaled corticosteroid fluticasone propionate given over six months suppressed EMT-related changes in large airways of COPD patients. This included marked reduction in Rbm fragmentation (Figure 1), EGFR epithelial expression, basal epithelial cell (Figure 2) and Rbm cell S100A4 expression (Figure 3) and Rbm cell MMP-9 staining in the active treatment arm (S. S. Sohal et al., 2014). Our data suggest a mechanism for the potential ICS preventative action against lung cancer in COPD. If this is true, it has huge implications for therapeutic and public health policy, since it is strongly suggested in the literature that patients on ICS have a marked decreased risk for lung cancer (S. S. Sohal et al., 2014).

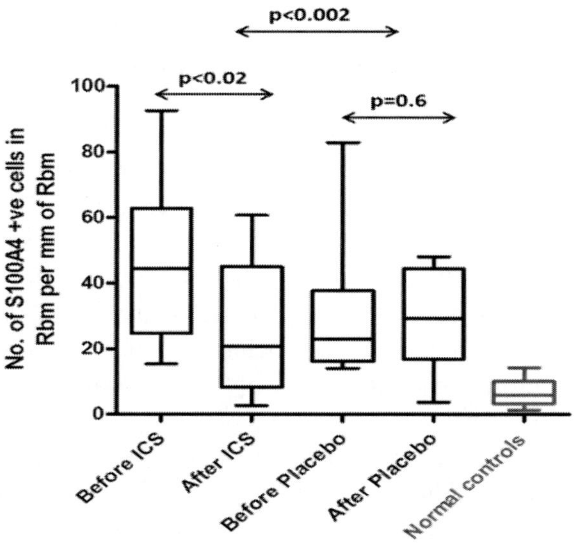

Figure 3. S100A4 expression in Reticular basement membrane (Rbm). Number of S100A4 positive cells in the Rbm before and after ICS versus before and after placebo, with normal control data for comparison. Changes over time with ICS were also significant compared to those on placebo ($p<0.002$). After treatment the active group was significantly different to normal controls ($p<0.004$). Data are represented as medians and ranges (S. S. Sohal et al., 2014).

There are a few recent studies reporting effects of other drugs on EMT. Milara et al. recently reported marked regression of EMT by roflumilast, a PDE4 inhibitor in bronchial epithelial cells in vitro by restoring cellular cyclic adenosine monophosphate (cAMP) levels (Milara et al., 2014). Wang and colleagues demonstrated increased urokinase-type plasminogen activator receptor (uPAR) expression, in the small airway epithelium of patients with COPD, related to active EMT, which could be blocked by antagonising uPAR (Q. Wang, Wang, Zhang, & Xiao, 2013). Other drugs which might have the potential to block EMT are nintedanib (multiple tyrosine kinases inhibitor) and pirfenidone (an anti-fibrotic and anti-inflammatory drug) (King et al., 2014; Richeldi et al., 2014). Nintedanib works mainly by inhibiting angiogenesis so it may have implications for EMT-Type-III where angiogenesis is prominent and on the other hand Pirfenidone is more anti-fibrotic in action so may have implications for EMT-Type-II (Singh Sukhwinder Sohal et al., 2014). These warrant further studies, as they may have both anti-fibrotic and anti-cancer effects by suppressing EMT.

Lung Cancer in COPD

The presence of COPD per-se increases the risk of developing lung cancer by 4-5 fold, when the smoking history is controlled for (Parimon et al., 2007). Further, up to 70% of lung cancer occurs in the context of COPD mainly in mild-moderate disease (P. J. Barnes & Adcock, 2011). This implies that mechanisms specific to the relatively early pathogenesis of COPD may be involved in development of lung cancer. Potential shared biological mechanisms in COPD and lung cancer includes: chronic inflammation, matrix degradation, cell proliferation and anti-apoptosis, abnormal wound repair and angiogenesis (I. A. Yang et al., 2011). All of these are associated with EMT, and especially EMT-Type-3 which is recognized as pro-malignant in other situations of potential epithelial malignancy (R. Kalluri & Weinberg, 2009). Indeed, the relationship between COPD pathology and carcinogenesis may reflect a more general paradigm of epithelial instability and cancer aetiology, bearing in mind that epithelial cancers make up 90% of all malignancies (Garber, 2008).

As mentioned above, a number of observational studies have demonstrated that ICS reduce local airway inflammation among patients with COPD (Hattotuwa et al., 2002; Reid et al., 2008). Animal models of smoking-induced COPD have demonstrated that glucocorticoids also strikingly inhibit development of smoking-related lung cancer (Greenberg et al., 2002; Wattenberg et al., 1997; Yao, Wang, Lemon, Lubet, & You, 2004). In human epidemiological research, a US veterans cohort study of 10,474 patients in primary care clinics found that use of ICS, albeit only at high doses, was associated with 50% decreased risk of lung cancer(Parimon et al., 2007). Similar findings were reported in ex-smokers with COPD (Kiri, Fabbri, Davis, & Soriano, 2009). Lee et al. reported ICS are associated with a reduced risk of lung cancer but not of laryngeal cancer (C. H. Lee et al., 2013). Similar observations were made by Schroedl and colleagues in COPD (Schroedl & Kalhan, 2012). Veronesi et al. reported in a randomized phase II trial of inhaled budesonide involving 202 former smokers with CT-detected lung nodules, a trend towards regression of nonsolid and partially solid nodules after budesonide treatment compared to the treatment arm (Veronesi et al., 2011). However, the TORCH study (Calverley et al., 2007) of ICS in severe COPD failed to show this effect, but it was not powered to pick this up, and was a study on severe COPD rather than mild to moderate COPD where one would expect such an effect of ICS to be most marked (P. J. Barnes & Adcock, 2011). This is a paradox that itself needs further understanding.

There is an essential need for both in vivo and in vitro human studies to understand the mechanistic link between COPD and airway cancer and how ICS and other drugs may affect it (van Gestel et al., 2009). We suggest that the effect we have shown of ICS in COPD on epithelial activation and aspects of EMT could be a key to such understanding, and could have major therapeutic implications (Miller & Keith, 2007; Nowrin, Sohal, Peterson, Patel, & Walters, 2014; S. Sohal, C. Ward, W. Danial, R. Wood-Baker, & E. Walters, 2013; Singh Sukhwinder Sohal et al., 2014; S. S. Sohal, A. Soltani et al., 2014).

Adverse Effects of Corticosteroids

Corticosteroids have a wide range of side effects which are well known and predictable and effects are commonly found in COPD patients who are undergoing inhaled and oral therapy.

In a case control study conducted it was found that there was a 70% rise in hospitalization in COPD patients (Ernst, Gonzalez, Brassard, & Suissa, 2007). Since then all major studies clinical studies have found a direct correlation between ICS dosage and pneumonia rates in COPD (Kardos, Wencker, Glaab, & Vogelmeier, 2007; Price et al., 2013; Wedzicha et al., 2008). For example the TORCH, a three year study reported an increase in hospitalisation up 19.6% and 18.3% in the fluticasone-containing arms compared with 13.3% with salmeterol and 12.3% with placebo (Calverley et al., 2007). Yawn et.al., (L. Finney et al.) found that low dose ICS had a 38% increase in pneumonia risk, while in medium and high dose corticosteroids treatment, had infection increased to 69% and 157% in risk respectively. The mechanism for this ICS related risk of pneumonia is not understood, but presumed to be due to suppression of innate immunity. In developing nations where malnutrition is a major problem, suppression of immune response by steroids can also lead to an increase in tuberculosis infection. In a case control study it was found that COPD patients administered oral doses of fluticasone of 1000μg per day were at a high risk of acquiring tuberculosis infections(Brassard, Suissa, Kezouh, & Ernst, 2011; Jick, Lieberman, Rahman, & Choi, 2006).

COPD patients on corticosteroids are also susceptible to other adverse effects such as reduced bone density leading to osteoporosis, early onset of diabetes, easy skin bruising, cataract, oral infection such as Oropharyngeal candidiasis and hoarseness. The risk of bone fractures and osteoporosis are highest among cigarette smoker especially in those who live sedentary lifestyles and have other co-morbidities. In fact osteoporosis have also been associated with milder stages of COPD and studies have shown ICS treatments further augment the risk of fractures (Graat-Verboom, van den Borne, Smeenk, Spruit, & Wouters, 2011; Lehouck, Boonen, Decramer, & Janssens, 2011). A large control study have showed that ICS treatment in COPD or asthma patients have a 34% higher risk of both new onset diabetes and diabetes progression and here too the study has shown dose dependent relation between ICS and increased diabetes susceptibility (Suissa, Dell'Aniello, & Ernst). Another adverse effect with ICS treatment is the increase risk of oral thrush and hoarseness, a systemic review found that COPD patients treated with ICS had increased risk of 2.98% for oral thrush and 2.2% for hoarseness (Sin, McAlister, Man, & Anthonisen, 2003). Also meta-analysis of forty seven primary studies that COPD patients had a risk ratio of 2.9% to acquire oropharyngeal candidiasis and hoarseness (I. A. Yang, Fong, Sim, Black, & Lasserson, 2007).

Conclusion

COPD is frequently said to be a “steroid-resistant” disease. This might be relatively so, but overall is untrue. COPD certainly cannot be “cured” with corticosteroids, but they are very much a central pillar of management in more severe disease, and positive effects are definitely demonstrable in both stable patients and in acute exacerbations. In this review we have detailed the evidence for especially ICS use and efficacy in COPD, and pointed towards where further research is urgently required. One of the most important aspects is the possibility of ICS being used for lung cancer prevention, possibly through an anti-EMT effect. This could have huge possible health importance, but generally receives very little attention.

Acknowledgments

Royal Hobart Hospital Research Foundation for funding.

References

Adcock, I. M., & Ito, K. (2005). Glucocorticoid pathways in chronic obstructive pulmonary disease therapy. *Proc Am Thorac Soc, 2*(4), 313-319; discussion 340-311. doi: 10.1513/pats.200504-035SR

Amir Soltani, H. K. M., Sukhwinder S. Sohal, David W. Reid, Steve Weston, Richard Wood-Baker, E. Haydn Walters (2012). The distinctive characteristics of bronchial reticular basement membrane and vessel remodelling in chronic obstructive pulmonary disease (COPD) versus asthma: they are not the same disease. *Histopathology, In press.*

Banerjee, D., Khair, O. A., & Honeybourne, D. (2004). Impact of sputum bacteria on airway inflammation and health status in clinical stable COPD. *Eur Respir J, 23*(5), 685-691.

Barbier, M., Agusti, A., & Alberti, S. (2008). Fluticasone propionate reduces bacterial airway epithelial invasion. *Eur Respir J, 32*(5), 1283-1288. doi: 10.1183/09031936.00020608

Barnes, P. J. (1998). Anti-inflammatory actions of glucocorticoids: molecular mechanisms. *Clin Sci (Lond), 94*(6), 557-572.

Barnes, P. J. (2006a). Corticosteroid effects on cell signalling. *Eur Respir J, 27*(2), 413-426. doi: 27/2/413 [pii]10.1183/09031936.06.00125404

Barnes, P. J. (2006b). How corticosteroids control inflammation: Quintiles Prize Lecture 2005. *Br J Pharmacol, 148*(3), 245-254. doi: 0706736 [pii]10.1038/sj.bjp.0706736

Barnes, P. J. (2006c). Reduced histone deacetylase in COPD: clinical implications. *Chest, 129*(1), 151-155. doi: 129/1/151 [pii]10.1378/chest.129.1.151

Barnes, P. J. (2006d). Transcription factors in airway diseases. *Lab Invest, 86*(9), 867-872. doi: 3700456 [pii]10.1038/labinvest.3700456

Barnes, P. J. (2008). Role of HDAC2 in the Pathophysiology of COPD. *Annu Rev Physiol*. doi: 10.1146/annurev.physiol.010908.163257

Barnes, P. J. (2009). Histone deacetylase-2 and airway disease. *Ther Adv Respir Dis, 3*(5), 235-243. doi: 1753465809348648 [pii]10.1177/1753465809348648

Barnes, P. J., & Adcock, I. M. (2011). Chronic obstructive pulmonary disease and lung cancer: a lethal association. *Am J Respir Crit Care Med, 184*(8), 866-867. doi: 184/8/866 [pii]10.1164/rccm.201108-1436ED

Barnes, P. J., Adcock, I. M., & Ito, K. (2005). Histone acetylation and deacetylation: importance in inflammatory lung diseases. *Eur Respir J, 25*(3), 552-563. doi: 25/3/552 [pii]10.1183/09031936.05.00117504

Barnes, P. J., & Karin, M. (1997). Nuclear factor-kappaB: a pivotal transcription factor in chronic inflammatory diseases. *N Engl J Med, 336*(15), 1066-1071.

Barnes, P. J., & Pedersen, S. (1993). Efficacy and Safety of Inhaled Corticosteroids in Asthma. *American Review of Respiratory Disease, 148*(4_pt_2), S1-S26. doi: 10.1164/ajrccm/148.4_Pt_2.S1

Brassard, P., Suissa, S., Kezouh, A., & Ernst, P. (2011). Inhaled Corticosteroids and Risk of Tuberculosis in Patients with Respiratory Diseases. *American Journal of Respiratory and Critical Care Medicine, 183*(5), 675-678. doi: 10.1164/rccm.201007-1099OC

Burge, P. S., Calverley, P. M., Jones, P. W., Spencer, S., Anderson, J. A., & Maslen, T. K. (2000). Randomised, double blind, placebo controlled study of fluticasone propionate in patients with moderate to severe chronic obstructive pulmonary disease: the ISOLDE trial. *Bmj, 320*(7245), 1297-1303.

Cabello, H., Torres, A., Celis, R., El-Ebiary, M., Puig de la Bellacasa, J., Xaubet A., González J., Agustí C., Soler N. (1997). Bacterial colonization of distal airways in healthy subjects and chronic lung disease: a bronchoscopic study. *Eur Respir J*, 10(5), 1137-1144.

Cabrera-Rubio, R., Garcia-Núñez, M., Setó, L., Antó, J. M., Moya, A., Monsó, E., & Mira, A. (2012). Microbiome Diversity in the Bronchial Tracts of Patients with Chronic Obstructive Pulmonary Disease. *Journal of Clinical Microbiology, 50*(11), 3562-3568. doi: 10.1128/jcm.00767-12

Calabrese, C., Bocchino, V., Vatrella, A., Marzo, C., Guarino, C., Mascitti, S., Tranfa CM., Cazzola M., Micheli P., Caputi M., Marsico, S. A. (2006). Evidence of angiogenesis in bronchial biopsies of smokers with and without airway obstruction. *Respir Med,* 100(8), 1415-1422. doi: 10.1016/j.rmed.2005.11.009

Calverley, P. M. (2004). Reducing the frequency and severity of exacerbations of chronic obstructive pulmonary disease. *Proc Am Thorac Soc, 1*(2), 121-124. doi: 10.1513/pats.2306031

Calverley P. M., Anderson J. A., Celli B., Ferguson G. T., Jenkins C., Jones P, W., Yates J. C, Vestbo J., TORCH investigators. (2007). Salmeterol and fluticasone propionate and survival in chronic obstructive pulmonary disease. *N Engl J Med,* 356(8), 775-789. doi: 356/8/775 [pii]10.1056/NEJMoa063070

Calverley, P. M., Boonsawat, W., Cseke, Z., Zhong, N., Peterson, S., & Olsson, H. (2003). Maintenance therapy with budesonide and formoterol in chronic obstructive pulmonary disease. *Eur Respir J, 22*(6), 912-919.

Calverley, P. M., Stockley, R. A., Seemungal, T. A., Hagan, G., Willits, L. R., Riley, J. H., & Wedzicha, J. A. (2011). Reported pneumonia in patients with COPD: findings from the INSPIRE study. *Chest, 139*(3), 505-512. doi: 10.1378/chest.09-2992

Celli BR., Thomas N. E., Anderson J. A., Ferguson G. T., Jenkins C. R., Jones P. W., Vestbo J., Knobil K., Yates J. C., Calverley P. M. (2008). Effect of pharmacotherapy on rate of decline of lung function in chronic obstructive pulmonary disease: results from the TORCH study. *Am J Respir Crit Care Med*, 178(4), 332-338. doi: 10.1164/rccm.200712-1869OC

Ceylan, E. (2006). Budesonide–formoterol (inhalation powder) in the treatment of COPD. *Int J Chron Obstruct Pulmon Dis, 1*(2), 115-122.

Chanez, P., Bourdin, A., Vachier, I., Godard, P., Bousquet, J., & Vignola, A. M. (2004). Effects of inhaled corticosteroids on pathology in asthma and chronic obstructive pulmonary disease. *Proc Am Thorac Soc, 1*(3), 184-190. doi: 1/3/184 [pii]10.1513/pats.200402-006MS

Chen, Y., Nickola, T. J., DiFronzo, N. L., Colberg-Poley, A. M., & Rose, M. C. (2006). Dexamethasone-mediated repression of MUC5AC gene expression in human lung epithelial cells. *Am J Respir Cell Mol Biol, 34*(3), 338-347. doi: 10.1165/rcmb.2005-0176OC

Cheng, S. L., Su, K. C., Wang, H. C., Perng, D. W., & Yang, P. C. (2014). Chronic obstructive pulmonary disease treated with inhaled medium- or high-dose corticosteroids: a prospective and randomized study focusing on clinical efficacy and the risk of pneumonia. *Drug Des Devel Ther, 8*, 601-607. doi: 10.2147/dddt.s63100

Chin C. L., Manzel L. J., Lehman E. E., Humlicek A. L., Shi L., Starner T. D., Denning G. M., Murphy T. F., Sethi S., Look DC. (2005). Haemophilus influenzae from patients with chronic obstructive pulmonary disease exacerbation induce more inflammation than colonizers. *Am J Respir Crit Care Med,* 172(1), 85-91. doi: 10.1164/rccm.200412-1687OC

Chung, K. F. (2005). The Role of Airway Smooth Muscle in the Pathogenesis of Airway Wall Remodeling in Chronic Obstructive Pulmonary Disease. *Proc Am Thorac Soc, 2*(4), 347-354. doi: 10.1513/pats.200504-028SR

Crim C., Calverley P. M., Anderson J. A., Celli B., Ferguson G. T., Jenkins C., Jones P. W., Willits L. R., Yates J. C., Vestbo J. (2009). Pneumonia risk in COPD patients receiving inhaled corticosteroids alone or in combination: TORCH study results. *Eur Respir J, 34*(3), 641-647. doi: 10.1183/09031936.00193908

de Ruijter, A. J., van Gennip, A. H., Caron, H. N., Kemp, S., & van Kuilenburg, A. B. (2003). Histone deacetylases (HDACs): characterization of the classical HDAC family. *Biochem J, 370*(Pt 3), 737-749. doi: 10.1042/BJ20021321BJ20021321 [pii]

Decramer, M., Janssens, W., & Miravitlles, M. (2012). Chronic obstructive pulmonary disease. *Lancet, 379*(9823), 1341-1351. doi: 10.1016/s0140-6736(11)60968-9

Degen, M., Goulet, S., Ferralli, J., Roth, M., Tamm, M., & Chiquet-Ehrismann, R. (2009). Opposite effect of fluticasone and salmeterol on fibronectin and tenascin-C expression in primary human lung fibroblasts. *Clin Exp Allergy, 39*(5), 688-699. doi: 10.1111/j.1365-2222.2009.03220.x

Dowling, R. B., Johnson, M., Cole, P. J., & Wilson, R. (1999). Effect of fluticasone propionate and salmeterol on Pseudomonas aeruginosa infection of the respiratory mucosa in vitro. *Eur Respir J, 14*(2), 363-369.

Edwards, M. R., Johnson, M. W., & Johnston, S. L. (2006). Combination therapy: Synergistic suppression of virus-induced chemokines in airway epithelial cells. *Am J Respir Cell Mol Biol, 34*(5), 616-624. doi: 10.1165/rcmb.2005-0385OC

Erb-Downward J. R., Thompson D. L., Han M. K., Freeman C. M., McCloskey L., Schmidt L. A., Young V. B., Toews G. B., Curtis J. L., Sundaram B., Martinez F. J., Huffnagle G. B. (2011). Analysis of the lung microbiome in the "healthy" smoker and in COPD. *PLoS One, 6*(2), e16384. doi: 10.1371/journal.pone.0016384

Ernst, P., Gonzalez, A. V., Brassard, P., & Suissa, S. (2007). Inhaled Corticosteroid Use in Chronic Obstructive Pulmonary Disease and the Risk of Hospitalization for Pneumonia. *American Journal of Respiratory and Critical Care Medicine, 176*(2), 162-166. doi: 10.1164/rccm.200611-1630OC

Ficner, R. (2009). Novel structural insights into class I and II histone deacetylases. *Curr Top Med Chem, 9*(3), 235-240.

Finney, L., Berry, M., Singanayagam, A., Elkin, S. L., Johnston, S. L., & Mallia, P. Inhaled corticosteroids and pneumonia in chronic obstructive pulmonary disease. *The Lancet Respiratory Medicine, 2*(11), 919-932. doi: 10.1016/S2213-2600(14)70169-9

Finney, L., Berry, M., Singanayagam, A., Elkin, S. L., Johnston, S. L., & Mallia, P. (2014). Inhaled corticosteroids and pneumonia in chronic obstructive pulmonary disease. *Lancet Respir Med, 2*(11), 919-932. doi: 10.1016/s2213-2600(14)70169-9

Gan, W. Q., Man, S. F., & Sin, D. D. (2005). Effects of inhaled corticosteroids on sputum cell counts in stable chronic obstructive pulmonary disease: a systematic review and a meta-analysis. *BMC Pulm Med, 5*, 3. doi: 10.1186/1471-2466-5-3

Garber, K. (2008). Epithelial-to-mesenchymal transition is important to metastasis, but questions remain. *J Natl Cancer Inst, 100*(4), 232-233, 239. doi: djn032 [pii]10.1093/jnci/djn032

Garcha, D. S., Thurston, S. J., Patel, A. R., Mackay, A. J., Goldring, J. J., Donaldson, G. C., . . . Wedzicha, J. A. (2012). Changes in prevalence and load of airway bacteria using quantitative PCR in stable and exacerbated COPD. *Thorax, 67*(12), 1075-1080. doi: 10.1136/thoraxjnl-2012-201924

Gizycki, M. J., Hattotuwa, K. L., Barnes, N., & Jeffery, P. K. (2002). Effects of fluticasone propionate on inflammatory cells in COPD: an ultrastructural examination of endobronchial biopsy tissue. *Thorax, 57*(9), 799-803.

Glare, E. M., Divjak, M., Bailey, M. J., & Walters, E. H. (2002). beta-Actin and GAPDH housekeeping gene expression in asthmatic airways is variable and not suitable for normalising mRNA levels. *Thorax, 57*(9), 765-770.

Goodin, D. S. (2014). Glucocorticoid treatment of multiple sclerosis. *Handb Clin Neurol, 122*, 455-464. doi: 10.1016/b978-0-444-52001-2.00020-0

Goulet, S., Bihl, M. P., Gambazzi, F., Tamm, M., & Roth, M. (2007). Opposite effect of corticosteroids and long-acting beta(2)-agonists on serum- and TGF-beta(1)-induced extracellular matrix deposition by primary human lung fibroblasts. *J Cell Physiol, 210*(1), 167-176. doi: 10.1002/jcp.20836

Graat-Verboom, L., van den Borne, B. E. E. M., Smeenk, F. W. J. M., Spruit, M. A., & Wouters, E. F. M. (2011). Osteoporosis in COPD outpatients based on bone mineral density and vertebral fractures. *Journal of Bone and Mineral Research, 26*(3), 561-568. doi: 10.1002/jbmr.257

Greenberg A. K., Hu J., Basu S., Hay J., Reibman J., Yie T. A., Tchou-Wong K. M., Rom W. N., Lee T. C. (2002). Glucocorticoids inhibit lung cancer cell growth through both the extracellular signal-related kinase pathway and cell cycle regulators. *Am J Respir Cell Mol Biol, 27*(3), 320-328.

Han M. K., Huang Y. J., Lipuma J. J., Boushey H. A., Boucher R. C., Cookson W. O., Curtis J. L., Erb-Downward J., Lynch S. V., Sethi S., Toews G. B., Young V. B., Wolfgang M. C., Huffnagle G. .B., Martinez F. J. (2012). Significance of the microbiome in obstructive lung disease. *Thorax, 67*(5), 456-463. doi: 10.1136/thoraxjnl-2011-201183

Hattotuwa, K. L., Gizycki, M. J., Ansari, T. W., Jeffery, P. K., & Barnes, N. C. (2002). The effects of inhaled fluticasone on airway inflammation in chronic obstructive pulmonary disease: a double-blind, placebo-controlled biopsy study. *Am J Respir Crit Care Med, 165*(12), 1592-1596.

Hogg, J. C., Chu, F., Utokaparch, S., Woods, R., Elliott, W. M., Buzatu, L., Cherniack R. M., Rogers R. M., Sciurba F. C., Coxson H. O., Pare, P. D. (2004). The Nature of Small-Airway Obstruction in Chronic Obstructive Pulmonary Disease. *N Engl J Med, 350*(26), 2645-2653. doi: 10.1056/NEJMoa032158

Innes, A. L., Carrington, S. D., Thornton, D. J., Kirkham, S., Rousseau, K., Dougherty, R. H., Raymond W. W., Caughey G. H., Muller S. J., Fahy, J. V. (2009). Ex vivo sputum analysis reveals impairment of protease-dependent mucus degradation by plasma proteins in acute asthma. *Am J Respir Crit Care Med, 180*(3), 203-210. doi: 10.1164/rccm.200807-1056OC

Ismaili, N., & Garabedian, M. J. (2004). Modulation of glucocorticoid receptor function via phosphorylation. *Ann N Y Acad Sci, 1024*, 86-101. doi: 10.1196/annals.1321.0071024/1/86 [pii]

Ito, K., Ito, M., Elliott, W. M., Cosio, B., Caramori, G., Kon, O. M., Barczyk A., Hayashi S., Adcock I. M., Hogg J. C., Barnes, P. J. (2005). Decreased histone deacetylase activity in chronic obstructive pulmonary disease. *N Engl J Med, 352*(19), 1967-1976. doi: 352/19/1967 [pii]10.1056/NEJMoa041892

Jeffery, P. K. (2001). Remodeling in asthma and chronic obstructive lung disease. *Am J Respir Crit Care Med, 164*(10 Pt 2), S28-38.

Jeffery, P. K. (2004). Remodeling and Inflammation of Bronchi in Asthma and Chronic Obstructive Pulmonary Disease. *Proc Am Thorac Soc, 1*(3), 176-183. doi: 10.1513/pats.200402-009MS

Jen, R., Rennard, S. I., & Sin, D. D. (2012). Effects of inhaled corticosteroids on airway inflammation in chronic obstructive pulmonary disease: a systematic review and meta-analysis. *Int J Chron Obstruct Pulmon Dis, 7*, 587-595. doi: 10.2147/copd.s32765

Jick, S. S., Lieberman, E. S., Rahman, M. U., & Choi, H. K. (2006). Glucocorticoid use, other associated factors, and the risk of tuberculosis. *Arthritis Care & Research, 55*(1), 19-26. doi: 10.1002/art.21705

John, M., Oltmanns, U., Binder, C., Meiners, S., Gellert, K., Chung, K. F., & Witt, C. (2004). Inhibition of chemokine production from human airway smooth muscle cells by fluticasone, budesonide and beclomethasone. *Pulm Pharmacol Ther, 17*(1), 41-47. doi: 10.1016/j.pupt.2003.09.002

Jones, P. W., Quirk, F. H., Baveystock, C. M., & Littlejohns, P. (1992). A self-complete measure of health status for chronic airflow limitation. The St. George's Respiratory Questionnaire. *Am Rev Respir Dis, 145*(6), 1321-1327. doi: 10.1164/ajrccm/145.6.1321

Kalluri, R. (2009). EMT: when epithelial cells decide to become mesenchymal-like cells. *J Clin Invest, 119*(6), 1417-1419. doi: 39675 [pii]10.1172/JCI39675

Kalluri, R., & Neilson, E. G. (2003). Epithelial-mesenchymal transition and its implications for fibrosis. *J. Clin. Invest., 112*(12), 1776-1784. doi: 10.1172/jci200320530

Kalluri, R., & Weinberg, R. A. (2009). The basics of epithelial-mesenchymal transition. *J Clin Invest, 119*(6), 1420-1428. doi: 39104 [pii] 10.1172/JCI39104

Kan, O. K., Matsumoto, K., Inoue, H., Fukuyama, S., Asai, Y., Watanabe, W., Kurokawa M., Araya J., Kuwano K., Nakanishi, Y. (2013). Corticosteroids plus long-acting beta2-agonists prevent double-stranded RNA-induced upregulation of B7-H1 on airway epithelium. *Int Arch Allergy Immunol, 160*(1), 27-36. doi: 10.1159/000338430

Kardos, P., Wencker, M., Glaab, T., & Vogelmeier, C. (2007). Impact of Salmeterol/Fluticasone Propionate versus Salmeterol on Exacerbations in Severe Chronic Obstructive Pulmonary Disease. *American Journal of Respiratory and Critical Care Medicine, 175*(2), 144-149. doi: 10.1164/rccm.200602-244OC

Kew, K. M., Mavergames, C., & Walters, J. A. (2013). Long-acting beta2-agonists for chronic obstructive pulmonary disease. *Cochrane Database Syst Rev, 10*, Cd010177. doi: 10.1002/14651858.CD010177.pub2

Kim, V., Rogers, T. J., & Criner, G. J. (2008). New concepts in the pathobiology of chronic obstructive pulmonary disease. *Proc Am Thorac Soc, 5*(4), 478-485.

King, T. E., Jr., Bradford, W. Z., Castro-Bernardini, S., Fagan, E. A., Glaspole, I., Glassberg, M. K., Gorina E., Hopkins P. M., Kardatzke D., Lancaster L., Lederer D. J., Nathan S. D., Pereira C. A., Sahn S. A.,

Sussman R., Swigris J. J., Noble, P. W. (2014). A phase 3 trial of pirfenidone in patients with idiopathic pulmonary fibrosis. *N Engl J Med, 370*(22), 2083-2092. doi: 10.1056/NEJMoa1402582

Kiri, V. A., Fabbri, L. M., Davis, K. J., & Soriano, J. B. (2009). Inhaled corticosteroids and risk of lung cancer among COPD patients who quit smoking. *Respir Med, 103*(1), 85-90. doi: S0954-6111(08)00275-8 [pii]10.1016/j.rmed.2008.07.024

Kranenburg, A. R., de Boer, W. I., Alagappan, V. K., Sterk, P. J., & Sharma, H. S. (2005). Enhanced bronchial expression of vascular endothelial growth factor and receptors (Flk-1 and Flt-1) in patients with chronic obstructive pulmonary disease. *Thorax, 60*(2), 106-113. doi: 60/2/106 [pii]10.1136/thx.2004.023986

Kunz L. I., Strebus J., Budulac S. E., Lapperre T. S., Sterk P. .J., Postma D. S., Mauad T., Timens W., Hiemstra P. S., GLUCOLD (Groningen Leiden Universities Corticosteroids in Obstructive Lung Disease) Study Group. (2013). Inhaled steroids modulate extracellular matrix composition in bronchial biopsies of COPD patients: a randomized, controlled trial. *PLoS One, 8*(5), e63430. doi: 10.1371/journal. pone.0063430

Lapperre T. S., Snoeck-Stroband J. B., Gosman M. M., Jansen D. F., van Schadewijk A., Thiadens H. A., Vonk J. M., Boezen H. M., Ten Hacken N. H., Sont J. K., Rabe K. F., Kerstjens H. A., Hiemstra P. S., Timens W., Postma D. S., Sterk P. J., Groningen Leiden Universities Corticosteroids in Obstructive Lung Disease Study Group. (2009). Effect of fluticasone with and without salmeterol on pulmonary outcomes in chronic obstructive pulmonary disease: a randomized trial. *Ann Intern Med, 151*(8), 517-527. doi: 151/8/517 [pii]

Lee, C. H., Hyun, M. K., Jang, E. J., Lee, N. R., Kim, K., & Yim, J. J. (2013). Inhaled corticosteroid use and risks of lung cancer and laryngeal cancer. *Respir Med, 107*(8), 1222-1233. doi: 10.1016/j.rmed.2012.12.002S0954-6111(12)00452-0 [pii]

Lee, S. Y., Kim, J. S., Lee, J. M., Kwon, S. S., Kim, K. H., Moon, H. S., Song JS., Park SH., Kim, Y. K. (2008). Inhaled corticosteroid prevents the thickening of airway smooth muscle in murine model of chronic asthma. *Pulmonary Pharmacology & Therapeutics, 21*(1), 14-19. doi: http://dx.doi.org/10. 1016/j.pupt.2006.10.009

Lehouck, A., Boonen, S., Decramer, M., & Janssens, W. (2011). COpd, bone metabolism, and osteoporosis. *Chest, 139*(3), 648-657. doi: 10.1378/chest.10-1427

Liesker, J. J., Ten Hacken, N. H., Zeinstra-Smith, M., Rutgers, S. R., Postma, D. S., & Timens, W. (2009). Reticular basement membrane in asthma and COPD: similar thickness, yet different composition. *Int J Chron Obstruct Pulmon Dis, 4*, 127-135.

MacRedmond, R. E., Greene, C. M., Dorscheid, D. R., McElvaney, N. G., & O'Neill, S. J. (2007). Epithelial expression of TLR4 is modulated in COPD and by steroids, salmeterol and cigarette smoke. *Respir Res, 8*, 84. doi: 10.1186/1465-9921-8-84

Marin A., Garcia-Aymerich J., Sauleda J., Belda J., Millares L., García-Núñez M., Serra I., Benet M., Agustí A., Antó J. M., Monsó E., PAC-COPD Study Group. (2012). Effect of bronchial colonisation on airway and systemic inflammation in stable COPD. *Copd, 9*(2), 121-130. doi: 10.3109/15412555.2011.636407

Milara, J., Peiro, T., Serrano, A., Guijarro, R., Zaragoza, C., Tenor, H., & Cortijo, J. (2014). Roflumilast N-oxide inhibits bronchial epithelial to mesenchymal transition induced by cigarette smoke in smokers with COPD. *Pulm Pharmacol Ther, 28*(2), 138-148. doi: 10.1016/ j.pupt.2014.02.001S1094-5539(14)00020-0 [pii]

Miller, Y. E., & Keith, R. L. (2007). Inhaled corticosteroids and lung cancer chemoprevention. *Am J Respir Crit Care Med, 175*(7), 636-637. doi: 175/7/636 [pii]10.1164/rccm.200701-087ED

Miravitlles M., Marín A., Monsó E., Vilà S., de la Roza C., Hervás R., Esquinas C., García M., Millares L., Morera J., Torres A. (2010). Colour of sputum is a marker for bacterial colonisation in chronic obstructive pulmonary disease. *Respir Res, 11*, 58. doi: 10.1186/1465-9921-11-58

Nannini, L. J., Poole, P., Milan, S. J., & Kesterton, A. (2013). Combined corticosteroid and long-acting beta(2)-agonist in one inhaler versus inhaled corticosteroids alone for chronic obstructive pulmonary disease. *Cochrane Database Syst Rev, 8*, Cd006826. doi: 10.1002/ 14651858.CD006826.pub2

Nowrin, K., Sohal, S. S., Peterson, G., Patel, R., & Walters, E. H. (2014). Epithelial-mesenchymal transition as a fundamental underlying pathogenic process in COPD airways: fibrosis, remodeling and cancer. *Expert Rev Respir Med*, 1-13. doi: 10.1586/17476348.2014.948853

Oltmanns U., Walters M., Sukkar M., Xie S., Issa R., Mitchell J., Johnson M., Chung K. F. (2008). Fluticasone, but not salmeterol, reduces cigarette smoke-induced production of interleukin-8 in human airway smooth muscle. *Pulm Pharmacol Ther, 21*(2), 292-297. doi: 10.1016/ j.pupt.2007.07.001

Osman, I. M., Godden, D. J., Friend, J. A., Legge, J. S., & Douglas, J. G. (1997). Quality of life and hospital re-admission in patients with chronic obstructive pulmonary disease. *Thorax, 52*(1), 67-71.

Parimon, T., Chien, J. W., Bryson, C. L., McDonell, M. B., Udris, E. M., & Au, D. H. (2007). Inhaled corticosteroids and risk of lung cancer among patients with chronic obstructive pulmonary disease. *Am J Respir Crit Care Med, 175*(7), 712-719. doi: 200608-1125OC [pii]10.1164/rccm.200608-1125OC

Patterson, C. M., Morrison, R. L., D'Souza, A., Teng, X. S., & Happel, K. I. (2012). Inhaled fluticasone propionate impairs pulmonary clearance of Klebsiella pneumoniae in mice. *Respir Res, 13*, 40. doi: 10.1186/1465-9921-13-40

Pauwels, R. A., Lofdahl, C. G., Laitinen, L. A., Schouten, J. P., Postma, D. S., Pride, N. B., & Ohlsson, S. V. (1999). Long-term treatment with inhaled budesonide in persons with mild chronic obstructive pulmonary disease who continue smoking. European Respiratory Society Study on Chronic Obstructive Pulmonary Disease. *N Engl J Med, 340*(25), 1948-1953.

Postma, D. S., & Timens, W. (2006). Remodeling in Asthma and Chronic Obstructive Pulmonary Disease. *Proc Am Thorac Soc, 3*(5), 434-439. doi: 10.1513/pats.200601-006AW

Pragman, A. A., Kim, H. B., Reilly, C. S., Wendt, C., & Isaacson, R. E. (2012). The Lung Microbiome in Moderate and Severe Chronic Obstructive Pulmonary Disease. *PLoS ONE, 7*(10), e47305. doi: 10.1371/journal.pone.0047305

Price, D., Yawn, B., Brusselle, G., & Rossi, A. (2013). Risk-to-benefit ratio of inhaled corticosteroids in patients with COPD. *Primary Care Respiratory Journal, 22*(1), 92-100. doi: 10.4104/pcrj.2012.00092

Reid, D. W., Wen, Y., Johns, D. P., Williams, T. J., Ward, C., & Walters, E. H. (2008). Bronchodilator reversibility, airway eosinophilia and anti-inflammatory effects of inhaled fluticasone in COPD are not related. *Respirology, 13*(6), 799-809. doi: RES1380 [pii]10.1111/j.1440-1843.2008.01380.x

Richeldi L., du Bois R. M., Raghu G., Azuma A., Brown K. K., Costabel U., Cottin V., Flaherty K. R., Hansell D. M., Inoue Y., Kim D. .S., Kolb M., Nicholson A. G., Noble P. W., Selman M., Taniguchi H., Brun M., Le Maulf F., Girard M., Stowasser S., Schlenker-Herceg R., Disse B., Collard HR., INPULSIS Trial Investigators (2014). Efficacy and safety of nintedanib in idiopathic pulmonary fibrosis. *N Engl J Med, 370*(22), 2071-2082. doi: 10.1056/NEJMoa1402584

Saglani S., Molyneux C., Gong H., Rogers A., Malmström K., Pelkonen A., Mäkelä M., Adelroth E., Bush A., Payne D. N., Jeffery P. K. (2006). Ultrastructure of the reticular basement membrane in asthmatic adults, children and infants. *Eur Respir J, 28*(3), 505-512. doi: 09031936.06.00056405 [pii]10.1183/09031936.06.00056 405

Sarir, H., Mortaz, E., Karimi, K., Johnson, M., Nijkamp, F. P., & Folkerts, G. (2007). Combination of fluticasone propionate and salmeterol potentiates the suppression of cigarette smoke-induced IL-8 production by macrophages. *Eur J Pharmacol, 571*(1), 55-61. doi: 10.1016/j.ejphar. 2007.05.034

Schroedl, C., & Kalhan, R. (2012). Incidence, treatment options, and outcomes of lung cancer in patients with chronic obstructive pulmonary disease. *Curr Opin Pulm Med, 18*(2), 131-137. doi: 10.1097/MCP.0b013 e32834f2080

Sethi, S. (2004). Bacteria in exacerbations of chronic obstructive pulmonary disease: phenomenon or epiphenomenon? *Proc Am Thorac Soc, 1*(2), 109-114. doi: 10.1513/pats.2306029

Shampo, M. A., & Kyle, R. A. Edward C. Kendall—Nobel Laureate. *Mayo Clinic Proceedings, 76*(12), 1188. doi: 10.4065/76.12.1188

Shampo, M. A., & Kyle, R. A. Philip S. Hench–1950 Nobel Laureate. *Mayo Clinic Proceedings, 76*(11), 1073. doi: 10.4065/76.11.1073

Shampo, M. A., Kyle, R. A., & Steensma, D. P. Tadeus Reichstein—Work on Hormones of the Adrenal Glands. *Mayo Clinic Proceedings, 88*(8), e85. doi: 10.1016/j.mayocp.2012.01.031

Siafakas, N. M., Antoniou, K. M., & Tzortzaki, E. G. (2007). Role of angiogenesis and vascular remodeling in chronic obstructive pulmonary disease. *Int J Chron Obstruct Pulmon Dis, 2*(4), 453-462.

Sin, D. D., McAlister, F. A., Man, S. F., & Anthonisen, N. R. (2003). Contemporary management of chronic obstructive pulmonary disease: scientific review. *Jama, 290*(17), 2301-2312. doi: 10.1001/jama. 290.17.2301

Singh, S., Amin, A. V., & Loke, Y. K. (2009). Long-term use of inhaled corticosteroids and the risk of pneumonia in chronic obstructive pulmonary disease: a meta-analysis. *Arch Intern Med, 169*(3), 219-229. doi: 10.1001/archinternmed.2008.550

Sohal, S., Ward, C., Danial, W., Wood-Baker, R., & Walters, E. (2013). Recent advances in understanding inflammation and remodeling in the airways in chronic obstructive pulmonary disease. *Expert Rev Respir Med, 7*(3), 275 - 288.

Sohal, S. S. (2014). Hot topics from the Assemblies. *Breathe, 10*(3), 263-264. doi: 10.1183/20734735.008014

Sohal, S. S., Mahmood, Q. M., & Walters, H. E. (2014). Clinical significance of epithelial mesenchymal transition (EMT) in chronic obstructive pulmonary disease (COPD): potential target for prevention of airway fibrosis and lung cancer. *Clinical and Translational Medicine, 3*(1), 33.

Sohal, S. S., Reid, D., Soltani, A., Ward, C., Weston, S., Muller, H. K., . . . Walters, E. H. (2010). Reticular basement membrane fragmentation and potential epithelial mesenchymal transition is exaggerated in the airways of smokers with chronic obstructive pulmonary disease. *Respirology, 15*(6), 930-938. doi: RES1808 [pii]10.1111/j.1440-1843.2010.01808.x

Sohal, S. S., Reid, D., Soltani, A., Ward, C., Weston, S., Muller, H. K., . . . Walters, E. H. (2011). Evaluation of epithelial mesenchymal transition in patients with chronic obstructive pulmonary disease. *Respir Res, 12*(1), 130. doi: 1465-9921-12-130 [pii]10.1186/1465-9921-12-130

Sohal, S. S., Reid D., Soltani A., Weston S., Muller HK., Wood-Baker R., Walters EH. (2010). Smoking cessation but not inhaled corticosteroids (ICS) can increase histone deacetylase 2 (HDAC2) in COPD. *European Respiratory Journal, Abstract book, 36*(54), 785S.

Sohal, S. S., Soltani, A., Reid, D., Ward, C., Wills, K. E., Muller, H. K., & Walters, E. H. (2014). A randomized controlled trial of inhaled corticosteroids (ICS) on markers of epithelial-mesenchymal transition (EMT) in large airway samples in COPD: an exploratory proof of concept study. *Int J Chron Obstruct Pulmon Dis, 9*, 533-542. doi: 10.2147/COPD.S63911copd-9-533 [pii]

Sohal, S. S., Soltani, A., Weston, S., Wood-Baker, R., & Walters, H. (2013). Intermediate filament vimentin and potential role in epithelial mesenchymal transition (EMT) *Vimentin concepts and molecular mechanisms* (pp. 37-61).

Sohal, S. S., & Walters, E. H. (2013a). Epithelial mesenchymal transition (EMT) in small airways of COPD patients. *Thorax, 68*(8), 783-784. doi: 10.1136/thoraxjnl-2013-203373thoraxjnl-2013-203373 [pii]

Sohal, S. S., & Walters, E. H. (2013b). Role of epithelial mesenchymal transition (EMT) in chronic obstructive pulmonary disease (COPD). *Respir Res, 14*(1), 120. doi: 1465-9921-14-120 [pii]10.1186/1465-9921-14-120

Sohal, S. S., Ward, C., Danial, W., Wood-Baker, R., & Walters, E. H. (2013). Recent advances in understanding inflammation and remodeling in the airways in chronic obstructive pulmonary disease. *Expert Rev Respir Med, 7*(3), 275-288. doi: 10.1586/ers.13.26

Sohal, S. S., Ward, C., & Walters, E. H. (2014). Importance of epithelial mesenchymal transition (EMT) in COPD and asthma. *Thorax*. doi: thoraxjnl-2014-205582 [pii]10.1136/thoraxjnl-2014-205582

Soler N., Torres A., Ewig S., Gonzalez J., Celis R., El-Ebiary M., Hernandez C., Rodriguez-Roisin R. (1998). Bronchial microbial patterns in severe exacerbations of chronic obstructive pulmonary disease (COPD) requiring mechanical ventilation. *Am J Respir Crit Care Med, 157*(5 Pt 1), 1498-1505. doi: 10.1164/ajrccm.157.5.9711044

Soltani, A., Muller, H., Sohal, S., Reid, D., Weston, S., Wood-Baker, R., & Walters, E. (2012). Distinctive characteristics of bronchial reticular basement membrane and vessel remodelling in chronic obstructive pulmonary disease (COPD) and in asthma: they are not the same disease. *Histopathology, 60*(6), 964 - 970.

Soltani, A., Reid, D. W., Sohal, S. S., Wood-Baker, R., Weston, S., Muller, H. K., & Walters, E. H. (2010). Basement membrane and vascular remodelling in smokers and chronic obstructive pulmonary disease: a cross-sectional study. *Respir Res, 11*(1), 105. doi: 1465-9921-11-105 [pii]10.1186/1465-9921-11-105

Soltani, A., Sohal, S., Reid, D., Weston, S., Wood-Baker, R., & Walters, E. (2012). Vessel-associated transforming growth factor-Beta1 (TGF-beta1) is increased in the bronchial reticular basement membrane in COPD and normal smokers. *PLoS One, 7*(6), e39736.

Soltani, A., Wood-Baker, R., Sohal, S. S., Muller, H. K., Reid, D., & Walters, E. H. (2012). Reticular Basement Membrane Vessels Are Increased in COPD Bronchial Mucosa by Both Factor VIII and Collagen IV Immunostaining and Are Hyperpermeable. *J Allergy (Cairo), 2012*, 958383. doi: 10.1155/2012/958383

Spahn, J. D., & Leung, D. Y. (1996). The role of glucocorticoids in the management of asthma. *Allergy Asthma Proc, 17*(6), 341-350.

Suissa, S., Dell'Aniello, S., & Ernst, P. (2012). Long-term natural history of chronic obstructive pulmonary disease: severe exacerbations and mortality. *Thorax, 67*(11), 957-963. doi: 10.1136/thoraxjnl-2011-201518

Szafranski W., Cukier A., Ramirez A., Menga G., Sansores R., Nahabedian S., Peterson S., Olsson H. (2003). Efficacy and safety of budesonide/ formoterol in the management of chronic obstructive pulmonary disease. *Eur Respir J, 21*(1), 74-81.

Sze M. A., Dimitriu P. A., Hayashi S., Elliott W. M., McDonough J. E., Gosselink J.V., Cooper J., Sin D. D., Mohn W. W., Hogg J. C. (2012). The lung tissue microbiome in chronic obstructive pulmonary disease. *Am J Respir Crit Care Med, 185*(10), 1073-1080. doi: 10.1164/rccm.201111-2075OC

Takami S., Mizuno T., Oyanagi T., Tadaki H., Suzuki T., Muramatsu K., Takizawa T., Arakawa H. (2012). Glucocorticoids inhibit MUC5AC production induced by transforming growth factor-alpha in human respiratory cells. *Allergol Int, 61*(3), 451-459. doi: 10.2332/allergolint.11-OA-0411

Thomas, M., Radwan, A., Stonham, C., & Marshall, S. (2014). COPD exacerbation frequency, pharmacotherapy and resource use: an observational study in UK primary care. *Copd, 11*(3), 300-309. doi: 10.3109/15412555.2013.841671

Triadafilopoulos, G. (2014). Glucocorticoid therapy for gastrointestinal diseases. *Expert Opin Drug Saf, 13*(5), 563-572. doi: 10.1517/14740338.2014.904852

Van Gestel Y. R., Hoeks S. E., Sin D. D., Hüzeir V., Stam H., Mertens F. W., van Domburg R. T., Bax J. J., Poldermans D. (2009). COPD and cancer mortality: the influence of statins. *Thorax, 64*(11), 963-967. doi: thx.2009. 116731 [pii]10.1136/thx.2009.116731

Vanacker, N. J., Palmans, E., Pauwels, R. A., & Kips, J. C. (2002). Effect of combining salmeterol and fluticasone on the progression of airway remodeling. *Am J Respir Crit Care Med, 166*(8), 1128-1134.

Veronesi G., Szabo E., Decensi A., Guerrieri-Gonzaga A., Bellomi M., Radice D., Ferretti S., Pelosi G., Lazzeroni M., Serrano D., Lippman S. M., Spaggiari L., Nardi-Pantoli A., Harari S., Varricchio C., Bonanni B. (2011). Randomized phase II trial of inhaled budesonide versus placebo in high-risk individuals with CT screen-detected lung nodules. *Cancer Prev Res (Phila), 4*(1), 34-42. doi: 10.1158/1940-6207.CAPR-10-01821940-6207.CAPR-10-0182 [pii]

Vestbo, J., Sorensen, T., Lange, P., Brix, A., Torre, P., & Viskum, K. (1999). Long-term effect of inhaled budesonide in mild and moderate chronic obstructive pulmonary disease: a randomised controlled trial. *Lancet, 353*(9167), 1819-1823.

Vlahos, R., Lee, K. S., Guida, E., Fernandes, D. J., Wilson, J. W., & Stewart, A. G. (2003). Differential inhibition of thrombin- and EGF-stimulated human cultured airway smooth muscle proliferation by glucocorticoids.

Pulm Pharmacol Ther, 16(3), 171-180. doi: 10.1016/s1094-5539(02)00183-9

Wang, P., Wang, X., Yang, X., Liu, Z., Wu, M., & Li, G. (2013). Budesonide suppresses pulmonary antibacterial host defense by down-regulating cathelicidin-related antimicrobial peptide in allergic inflammation mice and in lung epithelial cells. *BMC Immunol, 14*, 7. doi: 10.1186/1471-2172-14-7

Wang, Q., Wang, Y., Zhang, Y., & Xiao, W. (2013). The role of uPAR in epithelial-mesenchymal transition in small airway epithelium of patients with chronic obstructive pulmonary disease. *Respir Res, 14*, 67. doi: 10.1186/1465-9921-14-671465-9921-14-67 [pii]

Wattenberg, L. W., Wiedmann, T. S., Estensen, R. D., Zimmerman, C. L., Steele, V. E., & Kelloff, G. J. (1997). Chemoprevention of pulmonary carcinogenesis by aerosolized budesonide in female A/J mice. *Cancer Res, 57*(24), 5489-5492.

Wedzicha, J. A., Calverley, P. M. A., Seemungal, T. A., Hagan, G., Ansari, Z., & Stockley, R. A. (2008). The Prevention of Chronic Obstructive Pulmonary Disease Exacerbations by Salmeterol/Fluticasone Propionate or Tiotropium Bromide. *American Journal of Respiratory and Critical Care Medicine, 177*(1), 19-26. doi: 10.1164/rccm.200707-973OC

Weiner, J. M., Abramson, M. J., & Puy, R. M. (1998). Intranasal corticosteroids versus oral H1 receptor antagonists in allergic rhinitis: systematic review of randomised controlled trials. *Bmj, 317*(7173), 1624-1629.

Wilkinson, T. M., Hurst, J. R., Perera, W. R., Wilks, M., Donaldson, G. C., & Wedzicha, J. A. (2006). Effect of interactions between lower airway bacterial and rhinoviral infection in exacerbations of COPD. *Chest, 129*(2), 317-324. doi: 10.1378/chest.129.2.317

Wu, B., Li, P., Liu, Y., Lou, Z., Ding, Y., Shu, C., . . . Rao, Z. (2004). 3D structure of human FK506-binding protein 52: implications for the assembly of the glucocorticoid receptor/Hsp90/immunophilin heterocomplex. *Proc Natl Acad Sci U S A, 101*(22), 8348-8353. doi: 10.1073/pnas.03059691010305969101 [pii]

Yang, I. A., Clarke, M. S., Sim, E. H., & Fong, K. M. (2012). Inhaled corticosteroids for stable chronic obstructive pulmonary disease. *Cochrane Database Syst Rev, 7*, Cd002991. doi: 10.1002/14651858.CD002991.pub3

Yang, I. A., Fong, K. M., Sim, E. H., Black, P. N., & Lasserson, T. J. (2007). Inhaled corticosteroids for stable chronic obstructive pulmonary disease.

Cochrane Database Syst Rev(2), Cd002991. doi: 10.1002/1465 1858.CD002991.pub2

Yang I. A., Relan V., Wright C. M., Davidson M. R., Sriram K. B., Savarimuthu Francis S. M., Clarke B. E., Duhig E. E., Bowman R. V., Fong K. M. (2011). Common pathogenic mechanisms and pathways in the development of COPD and lung cancer. *Expert Opin Ther Targets, 15*(4), 439-456. doi: 10.1517/1472 8222.2011. 555400

Yang L., Yang X., Ji W., Deng J., Qiu F., Yang R., Fang W., Zhang L., Huang D., Xie C., Zhang H., Zhong N., Ran P., Zhou Y., Lu J. (2014). Effects of a functional variant c.353T>C in snail on risk of two contextual diseases. Chronic obstructive pulmonary disease and lung cancer. *Am J Respir Crit Care Med, 189*(2), 139-148. doi: 10.1164/rccm.201307-1355OC

Yao, R., Wang, Y., Lemon, W. J., Lubet, R. A., & You, M. (2004). Budesonide exerts its chemopreventive efficacy during mouse lung tumorigenesis by modulating gene expressions. *Oncogene, 23*(46), 7746-7752. doi: 10.1038/sj.onc.12079851207985 [pii]

Yildiz, F., Basyigit, I., Yildirim, E., Boyaci, H., & Ilgazli, A. (2004). Does addition of inhaled steroid to combined bronchodilator therapy affect health status in patients with COPD? *Respirology, 9*(3), 352-355. doi: 10.1111/j.1440-1843.2004.00618.x

Zeisberg, M., & Neilson, E. G. (2009). Biomarkers for epithelial-mesenchymal transitions. *J Clin Invest, 119*(6), 1429-1437. doi: 36183 [pii]10.1172 /JCI36183

Zhang, L., Prietsch, S. O., Mendes, A. P., Von Groll, A., Rocha, G. P., Carrion, L., & Da Silva, P. E. (2013). Inhaled corticosteroids increase the risk of oropharyngeal colonization by Streptococcus pneumoniae in children with asthma. *Respirology, 18*(2), 272-277. doi: 10.1111/j.1440-1843.2012.02280.x

Zhang, N., Truong-Tran, Q. A., Tancowny, B., Harris, K. E., & Schleimer, R. P. (2007). Glucocorticoids enhance or spare innate immunity: effects in airway epithelium are mediated by CCAAT/enhancer binding proteins. *J Immunol, 179*(1), 578-589.

In: Corticosteroids and Steroid Therapy
Editor: Carmen Adkins
ISBN: 978-1-63482-308-1
© 2015 Nova Science Publishers, Inc.

Chapter 2

Intranasal Steroid Treatment for Adenoids

Marco Berlucchi[1,*], Diego Barbieri[2] and Nader Nassif[1]
[1]Department of Pediatric Otorhinolaryngology, Spedali Civili, Brescia, Italy
[2]Department of Otorhinolaryngology, University of Brescia, Brescia, Italy

Abstract

Adenoid hypertrophy (AH) is a common childhood disease as associated with nasal obstruction with snoring, mouth breathing, hyponasal speech, rhinorrhea, and occasional abnormal facial development known as adenoid facies. By obstructing the rhinopharynx and nasopharyngeal orifice of the Eustachian tube, AH may also cause sinusitis and otitis media with effusion. Moreover, in the most serious cases, when associated with tonsillar hypertrophy, can cause obstructive sleep apnea syndrome. Treatment of AH in pediatric patients children depends on the degree of airway obstruction and related morbidity. Adenoidectomy has been traditionally considered to be definitive treatment for relief of upper airway obstruction and disease complicated by or attributable to AH. However, adenoidectomy has several pitfalls such as regrowth of adenoid after surgical removal, general complications

* Corresponding author: Marco Berlucchi, MD; Department of Pediatric Otorhinolaryngology, Spedali Civili, Piazza Spedali Civili 1, 25123 Brescia – Italy. Tel. +390303996226; Fax +390303996009; E-mail: marco.berlucchi@tin.it

(i.e., adverse anesthetic events and respiratory complications), and postoperative bleeding. A medical alternative to adenoidectomy is systemic steroid therapy, which leads to a prompt, temporary decrease in adenoid size, although chronic systemic administration is associated with serious adverse events.

An alternative to systemic steroids is the use of intranasal corticosteroids (INCS), which include beclomethasone dipropionate, budesonide, flunisolide, fluticasone propionate, and mometasone fluroate. INCS are associated with minimal systemic effects and they also have a substantially improved therapeutic index compared with intravenous and oral corticosteroids. The purpose of this review is to analyze the efficacy of each intranasal steroid and describe the spectrum of complications associated with their use.

Introduction

Adenoids, known as also pharyngeal or Lushka's tonsil, are a single, pyramid-shaped aggregation of lymphoid tissue located at the base of the roof and posterior wall of the nasopharynx. This lymphoid structure undergoes hypertrophy until 7 years of age, usually reaching a maximal size around the age of 4 years. Following this, it begins to atrophy until it almost invariably disappears in adulthood. The pharyngeal tonsil is a part of Waldeyer's ring, which is the first anatomic immunocompetent formation situated at the entrance of the aerodigestive tract and composed of the adenoid pad in addition to the palatine, tubal, and lingual tonsils [1].

Adenoid hypertrophy (AH) is probably the most frequent pediatric disease. An enlarged adenoid can occlude the choana (Figure 1), especially when sleeping in supine position, causing harmful effects such as nasal obstruction, snoring, mouth breathing, hyponasal speech, rhinorrhea, and abnormal facial development noted as adenoid facies. Such anomalies include an elongated face, prominent incisors, hypoplastic maxilla, short upper lip, elevated nostrils, and a high arched palate. Furthermore, when associated with tonsillar hypertrophy, obstructive sleep apnea syndrome (OSAS) can occur. Although a combination of structural and neuromuscular abnormalities contributes to occurrence of OSAS in children, the severity of OSAS is primarily related to adenoid and tonsillar size. The incidence of childhood OSAS has been estimated at 2–3% of all children, usually peaking between 2 and 8 years of age.

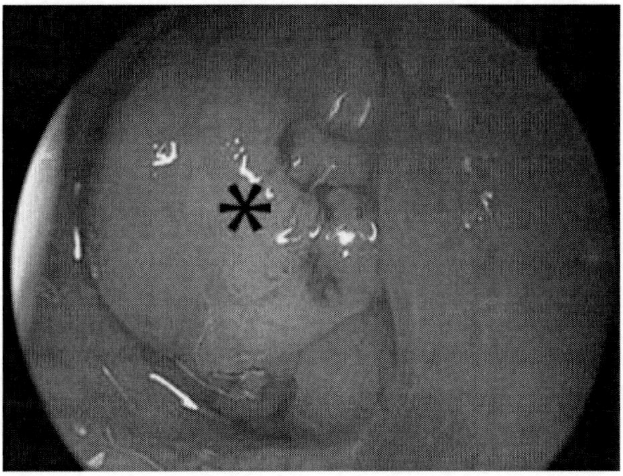

Figure 1. Adenoid hypertrophy (asterisk) obstruent totally the right choana.

Adenotonsillar hypertrophy is by far the major contributor to OSAS in children. Symptoms related to OSAS are intermittent sleep, sleepwalking, morning headaches, difficulty concentrating, sleepiness, enuresis, and slow feeding [2].

In the past, finger palpation, transoral mirror adenoid examination, and baseline lateral soft-tissue radiographs of the nasopharynx have been commonly used to assess adenoid size. At present, nasal endoscopy is considered the method of choice for the assessment of AH [3,4,5,6]. Improvement of endoscopic technology has led to the development of flexible and rigid endoscopes (Figures 2, 3) with small diameter (2.7 mm), which allows accurate nasal endoscopic examination with no complications. Rhinoscopy performed after positioning of cotton nasal pledgets soaked in topical anesthetic can be considered a thorough, safe, well-tolerated, and reproducible procedure for assessment of the adenoid pads (Figure 4).

Treatment of AH in children depends on the degree of airway obstruction and related morbidity. Adenoidectomy is considered to be definitive treatment for relief of upper airway obstruction.

However, this surgical therapy has several shortcomings: 1) Paulussen et al. hypothesized that the removal of adenoid lymphatic tissue could have a negative impact on the systemic immunologic system [7]; 2) regrowth of adenoid after surgical removal may occur in 10–20% of cases [8, 9, 10]; 3) adverse events related anesthesia (i.e., cardiac arrhythmia, malignant hyperthermia, vocal cord trauma and aspiration, dehydration) and respiratory complications such as upper airway edema, increased secretions, respiratory depression, and pulmonary edema may occur; and 4) postoperative bleeding is frequent. This latter complication is the most frequent and can range from mild bleeding that stops spontaneously to profuse bleeding demanding blood transfusion and surgical revision under general anesthesia.

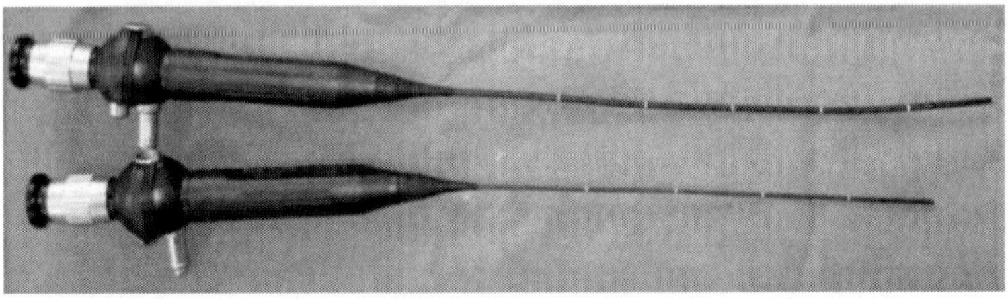

Figure 2. Flexible endoscopes with different diameter.

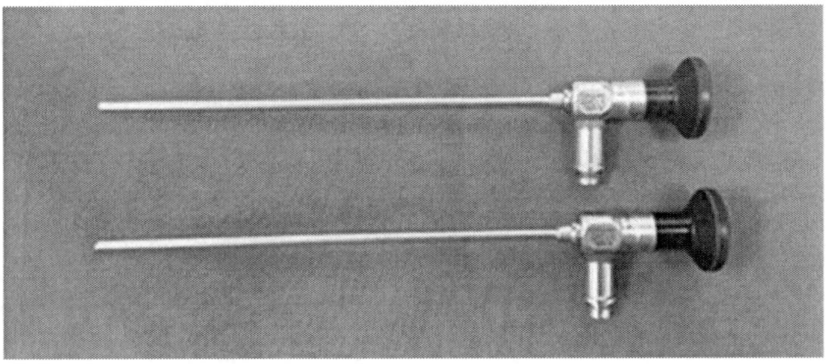

Figure 3. Rigid endoscopes with different diameter.

Recently, the incidence of hemorrhage after adenoidectomy has been reported to range from 0–0.49%, and is mainly related to adenoid remnants. Removal of these remnants under a second general anesthesia is the treatment of choice.

Medical alternatives to adenoidectomy are usually directed towards treatment of symptoms and concurrent infection with antibiotics, decongestants, and antihistamines. The use of systemic steroids produces a prompt, temporary decrease in adenoid size, but their chronic systemic administration is associated with serious adverse events [11].

An alternative to systemic steroids is intranasal corticosteroids (INCS) [1, 2, 12, 13, 14, 15, 16, 17, 18, 19, 20, 21, 22, 23, 24]. INCS include beclomethasone dipropionate (BDP), budesonide (BUD), flunisolide (FLU), fluticasone propionate (FP), and mometasone fluroate (MF). Treatment with INCS seems to decrease the size of adenoids through mechanisms that are still unclear. Reduction of adenoid size directly by a lympholytic effect, anti-inflammatory effects, and reducing the possibility of the adenoid acting as an infection reservoir are the main theories that have been hypothesized [12].

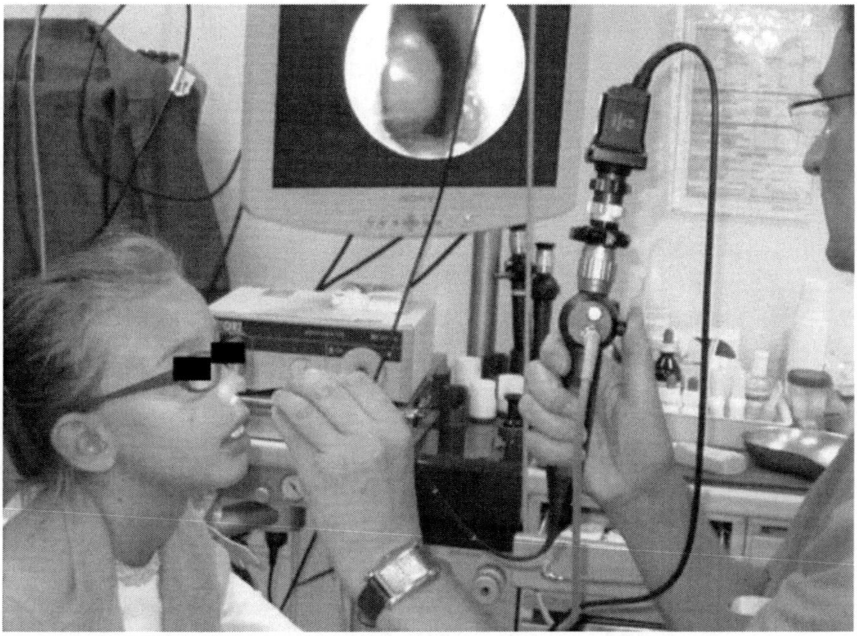

Figure 4. Execution of the nasal flexible endoscopy.

INCS, and especially the newer agents, do not cause clinically-significant suppression of the hypothalamic-pituitary-adrenal (HPA) axis at the recommended doses and are associated with a substantially improved therapeutic index compared with intravenous and oral corticosteroids. Moreover, INCS do not appear to be associated with reductions in either bone mineral density, osteoporosis [25, 26, 27, 28, 29, 30] or ocular changes [31, 32, 33, 34, 35, 36, 37, 38]. However, local side effects such as epistaxis, nasal crusting, dryness, throat irritation, burning, stinging, and nasal septum perforation can occur. [39, 40, 41, 42]. Among these, epistaxis is the most frequent.

The aim of this review is to assess the efficacy of each intranasal steroid on AH in children.

Intranasal Corticosteroids

Beclomethasone Dipropionate

BDP (Figure 5) was the first INCS evaluated, and was found to decrease obstructive nasal symptoms due to AH. Beclomethasone 17,21-dipropionate is a pro-drug with weak affinity for the glucocorticoid receptor, and is hydrolyzed via esterases to the active metabolite beclomethasone 17-monopropionate (B-17MP). Minor inactive metabolites, beclomethasone 21-monoproprionate (B-21-MP) and beclomethasone, are also formed. B-17-MP has a high bioavailability after nasal administration [43]. In 2001, Daley-Yates et al. demonstrated that the systemic exposure arises from the swallowed fraction, resulting in a similar total bioavailability to that found after oral dosing with the same formulation (≈40%). On the other hand, direct absorption in the nose was insignificant (<1%), and was limited by the short time available for dissolution of the poorly soluble compounds in the nose prior to clearance by the nasal cilia. Concerning possible adverse systemic effects, two studies in adult patients revealed that BDP did not affect adrenal function and did not cause suppression of the HPA axis [44, 45].

BDP was the first molecule assessed as a nonsurgical alternative for reduction of adenoid size. Three randomized controlled trials [12, 14, 15] were performed on a total of 111 children (aged 3 to 12 years), of whom 96 patients completed the respective study. In particular, two studies has a cross-over design (1 double-blind trial and 1 single-blind trial) [12, 15], whereas the remaining trial [14] had a placebo controlled, parallel-group design.

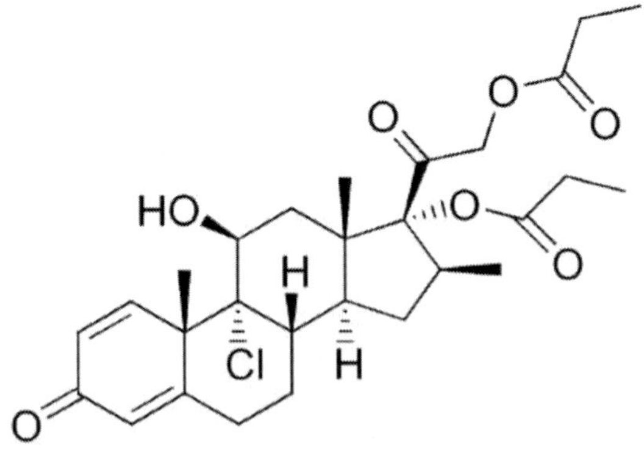

Figure 5. Chemical Structure of Beclomethasone diproprionate.

Inclusion criteria for the three trials required that each patient have AH [12, 14] or adenotonsillar hypertrophy [15]. AH was evaluated by nasal endoscopic examination in Demain and Goetz's study [12], by endoscopic procedure associated with lateral neck X-ray in the other two trials [14, 15]. Exclusion criteria utilized in these trials were the use of intranasal, topical, or systemic steroids within the last years, presence of an active upper respiratory infection within 2 weeks of entering the study, or history of chronic epistaxis and immunodeficiency. Furthermore, Demain and Goetz and Lepcha et al. excluded patients if they used intranasal medication within 2 weeks before entering the study [12, 14]. All the patients enrolled in Demain and Goetz's trial were individually randomized to receive either 4 weeks of intranasal aqueous BDP nasal spray (336 μg/day dispensed as two sprays in each nostril twice daily) followed by 4 weeks of placebo, or the two drugs in the reverse order. At the completion of the 8-week study, subjects continued in a 16-week open assessment of intranasal aqueous BDP nasal spray dispensed as a single spray (42 μg) in each nostril twice daily (168 μg/day). In Lepcha et al's trial, both the study and control groups received the drug (200 μg/day dispensed as one puff at a dose of 50 μg in each nostril twice daily) for 8 weeks, or placebo, respectively. In Criscuoli et al.'s trial, children were individually randomized to receive either 2 weeks of intranasal aqueous BDP nasal spray (400 μg/day dispensed as two sprays in each nostril twice daily) followed by 2 weeks of placebo or the two drugs in the reverse order. After this 4-week study, subjects continued in a 24-week open-label assessment of intranasal aqueous BDP nasal spray dispensed as 1 puff (50 μg) in each nostril twice daily.

In the three trials, improvement of obstructive nasal symptoms and adenoid size were employed as outcomes. The improvement of clinical picture was evaluated by different symptom scores, whereas decrease of adenoid pad was evaluated by nasal fiber optic examination in the three studies [12, 14, 15] associated with X-ray in Lepcha et al. and Criscuoli et al.'s trials [14, 15].

Regarding side effects, Demain and Goetz reported stinging in the 41%, epistaxis in 17%, and sneezing in 29% of all patients with a higher rate of stinging in the BDP group and greater sneezing in the placebo group. Lepcha et al. described no local complications. Criscuoli et al. reported epistaxis in 7% of patients who received BDP.

The Demain and Goetz study demonstrated the efficacy of topical nasal BDP in reducing either obstructive nasal symptoms in 51% of patients, or adenoid size in 17% of cases. These results were even more favorable at the end of the 16-week open continuation study, with an 82% reduction in mean nasal obstruction symptom score accompanied by a 29% mean reduction in adenoid size.

These results were confirmed in cross-over trial by Criscuoli et al. who observed that 45% of children showed clinical improvement after 2 weeks of intranasal steroid therapy. In addition, an additional 24-week treatment at a lower dosage of steroids was associated with significant clinical improvement at 52 and 100 weeks and reduction of adenotonsillectomy compared with children (55%) who had not responded to the initial 2-week therapy.

On the other hand, Lepcha et al. found no significant differences in nasal obstruction, nasal discharge, or snoring in children with AH between the BDP nasal spray and placebo groups. Those authors suggested that these results may be due to the fact that the study did not include children with allergy or atopy, and that the role of allergy in adenoid hypertrophy may be a possible explanation for the discrepancies in these results.

Thus, even if no significant improvements for the aforementioned reasons were observed in one study [14], BDP seems be effective in reducing adenoid size in the short term and in maintaining the good outcomes in the long term [12, 15].

Budesonide

Budesonide (Figure 6) is an INCS with a systemic bioavailability of about 34% [46]. To the best of our knowledge, there are no studies that have specifically investigated the role of BUD in reducing AH in children. On the other hand, a study showed that BUD may have a role in reducing adenoid size in children affected by OSAS.

In the study by Kheirandish-Gozal and Gozal [2] in 2007, 62 children (6−12 years) with OSAS were recruited into a double-blind, randomized, crossover trial. The treatment group was started on a 6-week course of intranasal topical BUD (32 μg per puff per nostril to both nostrils), whereas the control group received a placebo spray once daily (saline solution). After 2-week washout period, patients started a 6-week course with the agent that they were not receiving during the first phase of study. The authors concluded that 6 weeks of treatment with intranasal BUD effectively reduced the severity of mild OSAS and the magnitude of underlying adenoidal hypertrophy ($p<0.0001$). The effect persisted for at least 8 weeks after cessation of therapy. In summary, administration of BUD seems to favor improvement of the clinical picture in patients affected by AH.

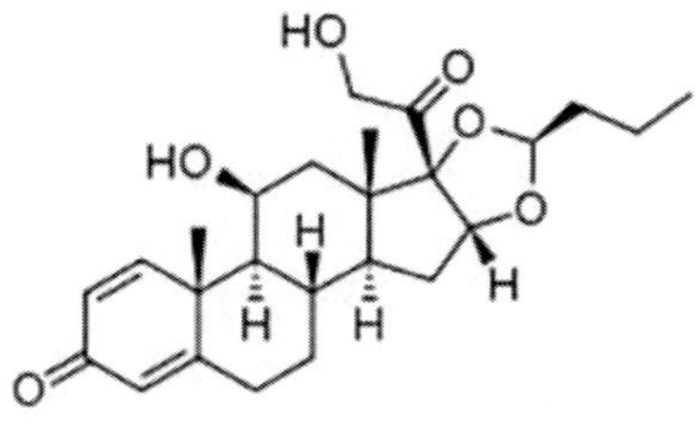

Figure 6. Chemical Structure of Budesonide.

Flunisolide

FLU (Figure 7) is an INCS characterized by high systemic bioavailability (≈ 49%). To the best of our knowledge, no study is available on the effect of intranasal FLU on function of the HPA axis [47].

In 2007, Ciprandi et al. [17] assessed the role of FLU in reducing AH in an 8-week, multicenter randomized, placebo-controlled, parallel-group trial. The 178 patients enrolled in the study were recruited from four Hospitals in Naples. All subjects complained of chronic nasal obstruction and were on the waiting list for adenoidectomy because of the presence of grade III or IV AH. Intranasal, topical, or systemic corticosteroids within the last year, intranasal medication within 2 weeks of entering the study, active upper respiratory infection within 2 weeks of entering the study, and/or history of chronic epistaxis, immunodeficiency or hypersensitivity to FLU were exclusion criteria.

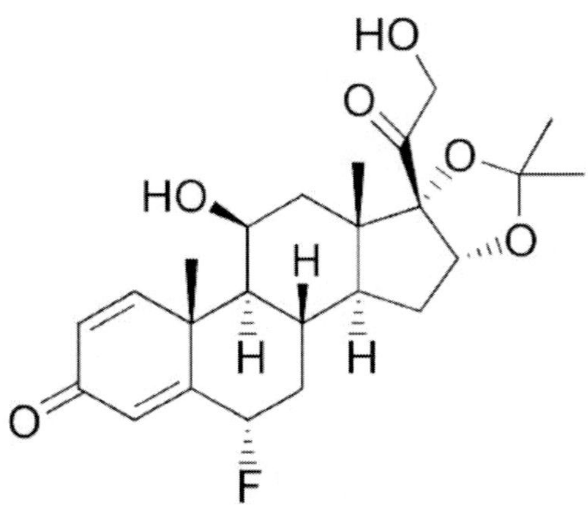

Figure 7. Chemical Structure of Flunisolide.

AH was evaluated by nasal fiber optic endoscopy and a symptom score was used to evaluate changes in symptomatology. Patients were randomly divided into two groups at a ratio of 3:1 (139 vs. 39) in order to obtain a large number of actively-treated patients. The study group was treated with FLU nasal drops (0.5 x kg body weight) for 8 weeks, and control one with normal saline solution. Variation in adenoid size was employed as the main outcome measure.

This 8-week trial showed that the use of intranasal FLU was associated with a significant reduction in size of adenoids in 72.6% ($p<0.02$) of children in comparison with isotonic saline solution. There was no significant difference between atopic and non-atopic children undergoing active treatment. No adverse events were reported.

In a subsequent randomized controlled study, Varricchio et al. [19] described the long-term effects of intranasal FLU on AH during 12-months of follow-up. Patients enrolled in the previous trial were re-evaluated by nasal fiber optic endoscopy at 6 and 12 months after suspension of treatment. It is interesting to note that most allergic children maintained the reduction in the size of AH ($p<0.05$). As suggested by authors, allergy is characterized by an inflammatory response and steroids are the most potent anti-inflammatory agents. In summary, two studies [17, 19] have documented the utility of FLU in reducing AH in children. Regarding short-term results, FLU seems able to reduce AH in all pediatric patients, whereas positive effects (i.e., stable results avoiding adenoidectomy) in a long-term follow up have been observed only in allergic children.

Fluticasone Propionate and Fluticasone Furoate

FP and FF (Figure 8a-b) are INCS characterized by low systemic bioavailability (<1%). At each nasal dose, approximately 30% of the drug is deposited in the nose where they bind to the glucocorticoid receptor. The remaining 70% of these intranasal steroids is swallowed and, about 99% of FP and 99.5% of FF, undergo first-pass hepatic metabolism [46].

The effects of FP and FF on the HPA axis in children have been investigated in several studies [48, 49, 50, 51, 52, 53], and no substantial alterations were found.

A randomized, triple-blind, placebo, controlled trial in 44 patients demonstrated the role of fluticasone in reducing the frequency of apnea and hypopnea in patients with OSAS. The drug was given as one spray of 50 μg per nostril twice daily for the first week and once daily for the subsequent 5 weeks [13]. However, fluticasone was not found to have any benefit in reducing AH.

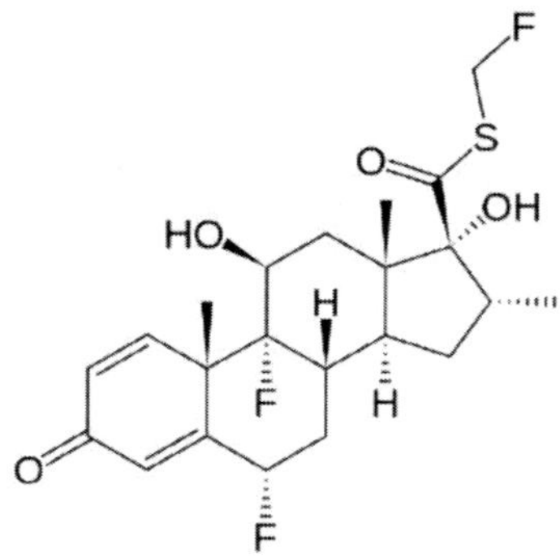

(a)

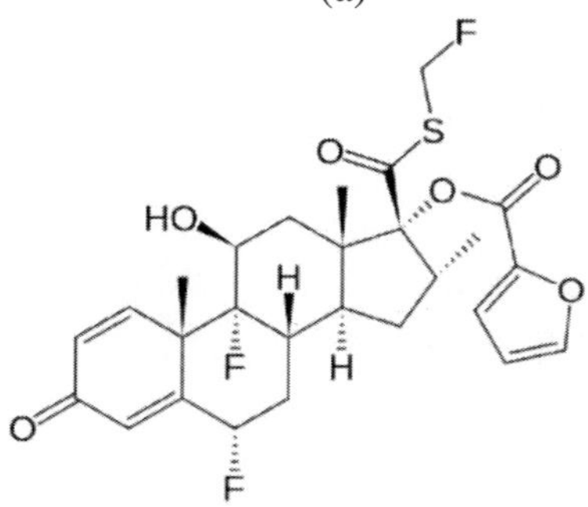

(b)

Figure 8. Chemical Structure of Fluticasone propionate (a) and Fluticasone furoate (b).

An 8-week randomized, placebo-controlled, parallel-group trial by Demirhan et al. in 2010 assessed the effectiveness of FP in reducing AH [20]. The 45 children enrolled in this study were indicated for adenotonsillectomy for recurrent tonsillitis and nasal symptoms related to AH lasting at least 6 months. Exclusion criteria were previous adenoidectomy, upper respiratory tract infection and/or allergic rhinitis, intranasal and/or systemic steroid therapy in the last year, intranasal medical treatment, chronic nose-bleeding, immunodeficiency, hypersensitivity, positive allergy and/or atopy against fluticasone, tonsillar hypertrophy, chronic otitis media with effusion and type B tympanogram, anatomic deformity of nose and/or sinonasal diseases, neurological diseases, and cardiovascular disease. AH was evaluated by nasal fiber optic endoscopy. A symptom scale was assessed before and after treatment. Patients were randomly divided into two groups to receive either fluticasone propionate nasal drops (400 μg/day) or saline solution for 8 weeks. Improvement of obstructive nasal symptoms and decrease of adenoid size were the main outcome measures. Intranasal FP was associated with good results in 76% of patients, thus avoiding surgical treatment. No adverse events were reported during the trial.

Recently, the efficacy of intranasal FF therapy has been evaluated on T-regulatory cells and other inflammation cytokines in adenoid tissues in children with OSAS [21]. In this randomized, prospective, open-label, parallel group trial in 24 children, the drug was given at the dose of 55 μg/nostril, once daily, for 2 weeks. The authors could not demonstrate a significant difference in adenoid size between the study and placebo groups. It was suggested that the reason for this discrepancy was that assessment of adenoid size was not performed before and after treatment with FF, but only after therapy and after adenoidectomy on the surgical specimen. On the other hand, the study showed a reduction in IL-6, a proinflammatory cytokine, in adenoid tissues treated with FF.

In summary, one study [20] has demonstrated the efficacy of FF in reducing AH, while another trial, which studied different endpoints, did not demonstrate any efficacy of FP in reducing AH [13].

Mometasone furoate

MF monohydrate (Figure 9) is an INCS characterized by low systemic bioavailability (<1%). The potency of MF is similar to FP, considered the most potent to date, and has almost undetectable systemic availability [54].

The drug does not cause any adverse tissue changes in the nasal mucosa of patients treated for long periods [40].

MF does not reach high systemic concentrations or cause clinically significant adverse effects. Results from pharmacokinetic studies in adults and children suggest that systemic exposure to MF after intranasal administration is negligible. This is probably because of the inherently low aqueous solubility of MF, which allows only a small fraction of the drug to cross the nasal mucosa and enter the bloodstream, and because a large amount of the administered drug is swallowed and undergoes extensive first-pass metabolism. The systemic availability of MF after topical administration is lower than other INCSs. There is no clinical evidence that MF nasal spray suppresses the function of the HPA axis, even when administered at clinically-relevant doses (100-200 µg/day) [54]. The effect of MF on the HPA axis has also been investigated in children [55, 56]. These trials all demonstrate that there are no relevant differences from baseline or placebo in any markers of adrenal suppression.

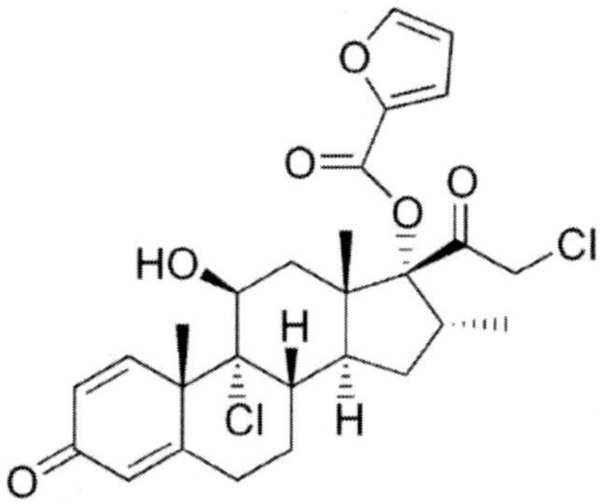

Figure 9. Chemical Structure of Mometasone furoate.

Two randomized controlled trials [1, 16] were performed on a total of 182 children (aged 3 to 15 years), of whom 179 were completers. The first [16] was a parallel open-label randomized controlled trial that evaluated the role of MF in treatment of pediatric patients with otitis media with effusion and/or AH. The second [1] was a two stage, double-blind placebo-controlled trial conducted with the only objective of demonstrating the efficacy of MF in reducing AH. In a later study [18], Berlucchi et al. assessed the utility of this INCS in the previous series after long-term follow-up.

Inclusion criteria for the two studies required that each patient have AH associated with or without OME [1, 16]. AH was evaluated by nasal endoscopic examination in both studies [1, 16]. Subjects in both studies were excluded if they had used intranasal, topical, or systemic steroids within the past 4 weeks, had an active upper respiratory infection within 2 weeks of entering the study, had history of immunodeficiency, hypersensitivity to mometasone furoate monohydrate or any systemic and local contraindication to corticosteroids or craniofacial anomaly [1, 16]. In one study additional exclusion criteria was concomitant tonsillar hypertrophy, positive history of allergy or atopy, nasal anatomic anomalies (e.g., nasal septum deviation) or sinonasal disease such as hypertrophy of inferior turbinates and/or nasal polyposis, neurologic disorders, cardiovascular diseases, history of epistaxis or intranasal, topical antibiotic treatment within the past 4 weeks [1].

The study group enrolled in Cengel and Akyol's trial (67 patients with AH, 34 with OME) received intranasal MF 100 µg/day, one spray in each nostril, once a day, for 6 weeks by the technique of neck flexion while dispensing from a vertically held bottle in order to direct the spray toward the posterior nasal cavity, whereas the control group (55 patients with AH, 29 with OME) was followed without any treatment. No other medication was allowed during the study in either group.

In the first stage of the trial by Berlucchi et al., children were randomly assigned to receive a single intranasal administration in each nostril of either MF aqueous nasal spray (50 µg) (group A) or a placebo saline solution nasal spray (group B) for 40 days. At the end of the first 40-day period, both groups were reassessed to evaluate the efficacy. The study-group patients, who showed improvement in clinical findings and decrease in adenoid pad size such that adenoidectomy could be avoided, were considered as responders. In the second stage, responders children underwent "maintenance therapy" for 3 months and were divided randomly into 2 subgroups: 1) group A1 received intranasal MF treatment on alternate days for the first 2 weeks per month, and 2) group A2 continued daily topical steroid treatment for the first 2 weeks for month.

In both trials, improvement of nasal obstruction symptoms and decrease of adenoid size were employed as outcome measures. The first outcome was evaluated by specific symptom scores, whereas second one it was assessed by flexible or rigid nasal endoscopy.

Regarding adverse events, Cengel and Akyol reported no adverse effects, whereas Berlucchi et al. described episodic epistaxis in 1 patient in the steroid group.

In the 6-week parallel open-label randomized controlled trial, Cengel and Akyol highlighted the positive role of MF in the short term in reducing AH and improving obstructive nasal symptoms. In fact, 45 patients (67.2%) in the steroid group showed a significant decrease in adenoid size ($p<0.001$) and significant improvement of the overall clinical picture ($p<0.001$) compared with the control group.

In the 3-month double-blind, placebo-controlled randomized cross-over trial, Berlucchi et al. showed that 77% of the study group obtained clinically relevant improvement after 40 days of nasal steroidal therapy. These patients, classified as responders, underwent two different maintenance therapies for 3 months. The daily use of MF for 2 weeks per month seemed to be the ideal maintenance schedule, and demonstrated the short-term efficacy of MF in reducing AH and related symptoms. Following this, Berlucchi et al. described the long-term follow-up of aforementioned pediatric cohort considered as "responders". Six children voluntarily suspended maintenance therapy, whereas the remaining 15 children continued the planned maintenance therapy, which included a single intranasal administration of MF aqueous nasal spray (50 μg) in each nostril daily for 2 weeks per month. Among these, 12 children, after a mean treatment period of 23 months (range 15−31 months), suspended treatment and did not undergo adenoidectomy.

The role of the MF in reducing AH in children was assessed in other two studies published in 2012 and 2013 [22, 23].

In the study by Rezende et al., 51 children (aged 4 to 8 years) with nasal obstructive symptoms due to AH were enrolled. In the first stage of the study, patients were instructed to use nasal saline douching in combination environmental prophylaxis for 40 days. After this initial period, topical mometasone furoate (100 μg/day) was given during the subsequent 40 days.

A semi-structured clinical questionnaire on nasal obstructive symptoms and nasal endoscopy were performed before starting treatment, after the first 40 days, and at the end of treatment with MF. The authors concluded that nasal saline douching significantly improved nasal symptoms without affecting adenoid dimension. In contrast, MF significantly decreased the size of adenoid tissue ($P<0.0001$), and led to a additional improvement in obstructive nasal symptoms.

The trial by Bitar et al. was a pilot study enrolling 19 children (aged 2 to 9 years) affected by AH. Patients received 100 μg of MF daily for 12 weeks. The results of treatment were compared with those of a matched control group (20 patients) who satisfied the inclusion criteria, but did not receive treatment. The authors concluded that MF nasal spray appears to be effective in treating children (aged 2-11 years) with adenoids obstructing more than 50% of the posterior choanae. Moreover, the authors stated that the effect of MF appears to be independent of the presence of mild intermittent allergic rhinitis and was not influenced by age of the patient or severity of symptoms. No complications or adverse effects were reported in the latter two studies.

Finally, the role of MF in reducing symptoms related to AH was evaluated in the adolescents. A 6-week prospective, double-blind, randomized, cross-over study, carried out on 28 subjects (12−18 years) with AH, showed significant improvement of the symptoms due to AH after intranasal application of MF (200 μg/day) compared with placebo without no reduction in adenoid size [24].

In summary, two randomized controlled trials [1, 16] demonstrated the efficacy of topical nasal MF in reducing adenoid size and in improving obstructive nasal symptoms in the short term in pediatric patients. In a consecutive study, Berlucchi et al. described the long-term efficacy of MF on adenoid tissue [18]. Furthermore, three studies [22, 23, 24] (2 in children and 1 in adolescents) confirmed the utility of MF in reducing AH and its correlated symptoms.

Conclusion

AH is a pediatric common disorder. To date, adenoidectomy is considered its definitive treatment. In the last decade, the use of INCS for AH obtained good results avoiding, thus, surgical therapy. Available INCS for this therapeutic procedure are different. The use of INCS can be considered as first-line therapy for AH. This treatment is conservative, safe, and well tolerated.

References

[1] Berlucchi, M; Salsi, D; Valetti, L; Parrinello, G; Nicolai, P. The role of mometasone furoate aqueous nasal spray in the treatment of adenoidal hypertrophy in the pediatric age group: preliminary results of a prospective, randomized study. *Pediatrics* 2007; 119, e1392-1397.

[2] Kheirandish-Gozal, L; Gozal, D. Intranasal budesonide treatment for children with mild obstructive sleep apnea syndrome. *Pediatrics* 2008; 122,e 149-155.

[3] Wang, DY; Clement, P; Kaufman, L: Derde, MP. Fiberoptic examination of the nasal cavity and nasopharynx in children. *Acta Otorhinolaryngol. Belg.* 1991; 45, 323-329.

[4] Kubba, H; Bingham, BJ. Endoscopy in the assessment of children with nasal obstruction. *J. Laryngol. Otol.* 2001; 115, 380-384.

[5] Cassano, P; Gelardi, M; Cassano, M; Fiorella, ML; Fiorella, R. Adenoid tissue rhinopharyngeal obstruction grading based on fiberendoscopic findings: a novel approach to therapeutic management. *Int. J. Pediatr. Otorhinolaryngol.* 2003; 67, 1303-1309.

[6] Chisholm, EJ; Lew-Gor, S; Hajioff, D; Caulfield, H. Adenoid size assessment: a comparison of palpation, nasendoscopy and mirror examination. *Clin. Otolaryngol.* 2005; 30, 39-41.

[7] Paulussen, C; Claes, J; Claes, G; Jorissen, M. Adenoids and tonsils, indications for surgery and immunological consequences of surgery. *Acta Otorhinolaryngol. Belg.* 2000; 54, 403-408.

[8] Rasmussen, N. Complications of tonsillectomy and adenoidectomy. *Otolaryngol. Clin. North Am.* 1987; 20, 383-390.

[9] Kovaleva, LM. Repeated adenoidectomy and prevention of the recurrence of adenoid hypertrophy. *Vestn. Otorinolaringol.* 1994; (1), 18-21.

[10] Buchinsky, FJ; Lowry, MA; Isaacson, G. Do adenoids regrow after excision? *Otolaryngol. Head Neck Surg.* 2000; 123, 576-581.

[11] Reilly, JS. Tonsillar and adenoid airway obstruction: modes of treatment in children. *Int. Anesthesiol. Clin.* 1988; 26, 54-57.

[12] Demain, JG; Goetz, DW. Pediatric adenoidal hypertrophy and nasal airway obstruction: reduction with aqueous nasal beclomethasone. *Pediatrics* 1995; 95, 355-364.

[13] Brouillette, RT, Manoukian, JJ; Ducharme, FM; Oudjhane, K; Earle, LG; Ladan, S et al. Efficacy of fluticasone nasal spray for pediatric obstructive sleep apnea. *J. Pediatr.* 2001; 138, 838-844.

[14] Lepcha, A; Kurien, M; Job, A; Jeyaseelan, L; Thomas, K. Chronic adenoid hypertrophy in children - is steroid nasal spray beneficial? Indian J. Otolaryngol. *Head Neck Surg.* 2002; 54, 280-284.

[15] Criscuoli, G; D'Amora, S; Ripa, G; Cinquegrana, G; Mansi, N; Impagliazzo, N et al. Frequency of surgery among children who have adenotonsillar hypertrophy and improve after treatment with nasal beclomethasone. *Pediatrics* 2003; 111, e236-238.

[16] Cengel, S, Akyol, MU. The role of topical nasal steroids in the treatment of children with otitis media with effusion and/or adenoid hypertrophy. *Int. J. Pediatr. Otorhinolaryngol.* 2006; 70, 639-645.

[17] Ciprandi, G; Varricchio, A; Capasso, M; Varricchio, AM; De Lucia, A; Ascione, E et al. Intranasal flunisolide treatment in children with adenoidal hypertrophy. *Int. J. Immunopathol. Pharmacol.* 2007; 20, 833-836.

[18] Berlucchi, M; Valetti, L; Parrinello, G; Nicolai, P. Long-term follow-up of children undergoing topical intranasal steroid therapy for adenoidal hypertrophy. *Int. J. Pediatr. Otorhinolaryngol.* 2008; 72, 1171-1175.

[19] Varricchio, A; Tortoriello, G; Capasso, M; De Lucia, A; Marchisio, P; Varricchio, AM et al. Prevention of surgery in children with adenoidal hypertrophy treated with intranasal flunisolide: a 12-month follow-up. *J. Biol. Regul. Homeost. Agents* 2009; 23, 95-101.

[20] Demirhan, H; Aksoy, F; Ozturan, O; Yildirim, YS; Veyseller, B. Medical treatment of adenoid hypertrophy with "fluticasone propionate nasal drops". *Int. J. Pediatr. Otorhinolaryngol.* 2010; 74, 773-776.

[21] Esteitie, R; deTineo, M; Naclerio, RM; Baroody, FM. Effect of the addition of montelukast to fluticasone propionate for the treatment of perennial allergic rhinitis. *Ann. Allergy Asthma Immunol.* 2010; 105, 155-161.

[22] Rezende, RM, Silveira, F; Barbosa, AP; Menezes, UP; Ferriani, VP; Rezende, PH et al. Objective reduction in adenoid tissue after mometasone furoate treatment. *Int. J. Pediatr. Otorhinolaryngol.* 2012; 76, 829-831.

[23] Bitar, MA; Mahfoud, L; Nassar, J; Dana, R. Exploring the characteristics of children with obstructive adenoid responding to mometasone fuorate monohydrate: preliminary results. *Eur. Arch, Otorhinolaryngol.* 2013; 270, 931-937.

[24] Yilmaz, HB; Celebi, S; Sahin-Yilmaz, A; Oysu, C. The role of mometasone furoate nasal spray in the treatment of adenoidal hypertrophy in the adolescents: a prospective, randomized, cross-over study. *Eur. Arch. Otorhinolaryngol.* 2013; 270, 2657-2661.

[25] Martinati, LC; Sette, L; Chiocca, E; Zaninotto, M; Plebani, M; Boner, AL. Effect of beclomethasone dipropionate nasal aerosol on serum markers of bone metabolism in children with seasonal allergic rhinitis. *Clin. Exp. Allergy* 1993; 23, 986-991.

[26] Howland, WC 3rd. Fluticasone propionate: topical or systemic effects? *Clin. Exp. Allergy* 1996; 26 Suppl 3, 18-22.

[27] Howland, WC 3rd; Dockhorn, R; Gillman, S; Gross, GN; Hille, D; Simpson, B; Furst, JA; Feiss, G; Smith, JA. A comparison of effects of triamcinolone acetonide aqueous nasal spray, oral prednisone, and placebo on adrenocortical function in male patients with allergic rhinitis. *J. Allergy Clin. Immunol.* 1996; 98, 32-38.

[28] Wilson, AM; Sims, EJ; McFarlane, LC; Lipworth, BJ. Effects of intranasal corticosteroids on adrenal, bone, and blood markers of systemic activity in allergic rhinitis. *J. Allergy Clin. Immunol.* 1998; 102, 598-604.

[29] Cave, A; Arlett, P; Lee, E. Inhaled and nasal corticosteroids: factors affecting the risks of systemic adverse effects. *Pharmacol. Ther.* 1999; 83, 153-179.

[30] Lee, DK; Robb, FM; Sims, EJ; Currie, GP; McFarlane, LC; Lipworth, BJ. Systemic bioactivity of intranasal triamcinolone and mometasone in perennial allergic rhinitis. *Br. J. Clin. Pharmacol.* 2003; 55, 310-313.

[31] Oztürk, F; Yücetürk, AV; Kurt, E; Unlü, HH; Ilker, SS. Evaluation of intraocular pressure and cataract formation following the long-term use of nasal corticosteroids. *Ear Nose Throat J.* 1998; 77, 846-8, 850-1.

[32] Dibildox, J. Safety and efficacy of mometasone furoate aqueous nasal spray in children with allergic rhinitis: results of recent clinical trials. *J. Allergy Clin. Immunol.* 2001; 108 Suppl 1, S54-58.

[33] Bross-Soriano, D; Hanenberg-Milver, C; Schimelmitz-Idi, J; Arrieta-Gomez, JR; Astorga del Toro, R; Bravo-Escobar, G. Effects of three nasal topical steroids in the intraocular pressure compartment. *Otolaryngol. Head Neck Surg.* 2004; 130, 187-191.

[34] Behbehani, AH; Owayed, AF; Hijazi, ZM; Eslah, EA; Al-Jazzaf, AM. Cataract and ocular hypertension in children on inhaled corticosteroid therapy. *J. Pediatr. Ophthalmol. Strabismus* 2005; 42, 23-27.

[35] Ernst, P; Baltzan, M; Deschênes, J; Suissa, S. Low-dose inhaled and nasal corticosteroid use and the risk of cataracts. *Eur. Respir. J.* 2006; 27, 1168-1174.

[36] Rosenblut, A; Bardin, PG; Muller, B; Faris, MA; Wu, WW; Caldwell, MF; Fokkens, WJ. Long-term safety of fluticasone furoate nasal spray in adults and adolescents with perennial allergic rhinitis. *Allergy* 2007; 62, 1071-1077.

[37] Máspero, JF; Rosenblut, A; Finn, A Jr; Lim J; Wu, W; Philpot, E. Safety and efficacy of fluticasone furoate in pediatric patients with perennial allergic rhinitis. *Otolaryngol. Head Neck Surg.* 2008; 138, 30-37.

[38] Ratner, PH; Meltzer, EO; Teper, A. Mometasone furoate nasal spray is safe and effective for 1-year treatment of children with perennial allergic rhinitis. *Int. J. Pediatr. Otorhinolaryngol.* 2009; 73, 651-657.

[39] Holm, AF; Fokkens, WJ; Godthelp, T; Mulder, PG; Vroom, TM; Rijntjes, E. A 1-year placebo-controlled study of intranasal fluticasone propionate aqueous nasal spray in patients with perennial allergic rhinitis: a safety and biopsy study. *Clin. Otolaryngol. Allied Sci.* 1998; 23, 69-73.

[40] Minshall, E; Ghaffar, O; Cameron, L; O'Brien, F; Quinn, H; Rowe-Jones, J et al. Assessment by nasal biopsy of long-term use of

mometasone furoate aqueous nasal spray (Nasonex) in the treatment of perennial rhinitis. Otolaryngol. *Head Neck Surg.* 1998; 118, 648-654.

[41] Klossek, JM; Laliberté, F; Laliberté, MF; Mounedji, N; Bousquet, J. Local safety of intranasal triamcinolone acetonide: clinical and histological aspects of nasal mucosa in the long-term treatment of perennial allergic rhinitis. *Rhinology* 2001; 39, 17-22.

[42] Scadding, G; Erkan, AN; Chau, H; Maskell, S. Audit of nasal steroid use and effectiveness in a rhinitis clinic. *Expert Rev. Pharmacoecon. Outcomes Res.* 2010; 10, 87-90.

[43] Daley-Yates, PT; Price, AC; Sisson, JR; Pereira, A; Dallow, N. Beclomethasone dipropionate: absolute bioavailability, pharmacokinetics and metabolism following intravenous, oral, intranasal and inhaled administration in man. *Br. J. Clin. Pharmacol.* 2001; 51, 400-409.

[44] Brannan, MD; Herron, JM; Reidenberg, P; Affrime, MB. Lack of hypothalamic-pituitary-adrenal axis suppression with once-daily or twice-daily beclomethasone dipropionate aqueous nasal spray administered to patients with allergic rhinitis. *Clin. Ther.* 1995; 17, 637-647.

[45] Giger, R; Pasche, P; Cheseaux, C; Cantini, L; Rossetti, A; Landis, BN et al. Comparison of once- versus twice-daily use of beclomethasone dipropionate aqueous nasal spray in the treatment of allergic and non-allergic chronic rhinosinusitis. *Eur. Arch. Otorhinolaryngol.* 2003; 260, 135-140.

[46] Sastre, J; Mosges, R. Local and systemic safety of intranasal corticosteroids. *J. Investig. Allergol. Clin. Immunol.* 2012; 22, 1-12.

[47] Bruni, FM; De Luca, G; Venturoli, V; Boner, AL. Intranasal corticosteroids and adrenal suppression. *Neuroimmunomodulation.* 2009; 16, 353-362.

[48] Grossman, J; Banov, C; Bronsky, EA; Nathan RA; Pearlman D; Winder JA et al. Fluticasone propionate aqueous nasal spray is safe and effective for children with seasonal allergic rhinitis. *Pediatrics* 1993; 92, 594-599.

[49] Treatment of seasonal allergic rhinitis with once-daily intranasal fluticasone propionate therapy in children. Fluticasone Propionate Collaborative Pediatric Working Group. *J. Pedia*tr. 1994; 125, 628-634.

[50] Ngamphaiboon, J; Thepchatri, A; Chatchatee, P; Chumdermpadetsuk, S. Fluticasone propionate aqueous nasal spray treatment for perennial allergic rhinitis in children. *Ann. Allergy Asthma Immunol.* 1997; 78, 479-484.

[51] Galant, SP; Melamed, IR; Nayak, AS; Blake, KV; Prillaman, BA; Reed, KD et al. Lack of effect of fluticasone propionate aqueous nasal spray on the hypothalamic-pituitary-adrenal axis in 2- and 3-year-old patients. *Pediatrics* 2003; 112, 96-100.

[52] Meltzer, EO; Tripathy, I; Máspero, JF; Wu, W; Philpot, E. Safety and tolerability of fluticasone furoate nasal spray once daily in paediatric patients aged 6-11 years with allergic rhinitis: subanalysis of three randomized, double-blind, placebo-controlled, multicentre studies. *Clin. Drug Investig*. 2009; 29, 79-86.

[53] Tripathy, I; Levy, A; Ratner, P; Clements, D; Wu, W; Philpot, E. HPA axis safety of fluticasone furoate nasal spray once daily in children with perennial allergic rhinitis. *Pediatr. Allergy Immunol.* 2009; 20, 287-294.

[54] Onrust, SV; Lamb, HM. Mometasone furoate. A review of its intranasal use in allergic rhinitis. *Drugs* 1998; 56, 725-745.

[55] Boner, AL. Effects of intranasal corticosteroids on the hypothalamic-pituitary-adrenal axis in children. *J. Allergy Clin. Immunol.* 2001; 108 Suppl 1, S32-39.

[56] Schenkel, EJ; Skoner, DP; Bronsky, EA; Miller, SD; Pearlman, DS; Rooklin, A et al. Absence of growth retardation in children with perennial allergic rhinitis after one year of treatment with mometasone furoate aqueous nasal spray. *Pediatrics* 2000; 105, E22.

In: Corticosteroids and Steroid Therapy
Editor: Carmen Adkins
ISBN: 978-1-63482-308-1
© 2015 Nova Science Publishers, Inc.

Chapter 3

The Role of Steroids in the Management of Chronic Subdural Hematoma: Principles and Clinical Considerations

Julio Plata Bello
Hospital Universitario de Canarias (Department of Neurosurgery), S/C de Tenerife, Spain

Abstract

Chronic subdural hematoma (CSDH) is a common condition in the elderly population and one of the most frequent lesions encountered in neurosurgical departments.

Mild head trauma is reported in most cases, but the pathophysiology of CSDH is still a matter of debate. Several data support the role of inflammatory related factors in the pathogenesis of the lesion, thus CSDH is considered a chronic self-perpetuating inflammatory process involving the dura matter.

Surgical treatment is the most common procedure for this type of lesion and it has proved to be effective. However, there is a large amount of data supporting the use of steroids in the management of CSDH. This data is essentially based on the inflammatory processes that have been postulated as underlying CSDH development.

The aim of this chapter is to describe the current role of steroids in the management of CSDH based on the pathophysiological processes that have been postulated as underlying CSDH development.

Introduction

Chronic subdural hematoma (CSDH) is one of the most common diseases seen in routine neurosurgical care. CDSH consists of a slow progressive collection of fluid in the subdural space (i.e., between the surface of the brain and the dura matter). This space is normally a virtual space but some pathological conditions can cause a build-up of material in this space (e.g., acute subdural hematoma, subdural empyema, pneumoencephalus). The fluid content in cases of CSDH is a combination of cephalous-spinal fluid (CSF) and blood degradation products.

The incidence of CSDH is around 13.5 cases per 100,000 individuals per year in the general population. This incidence is even higher when only patients over 65 years of age are considered (estimated incidence of 58.1 per 100,000).

There are some risk factors facilitating the development of subdural collections. They include chronic alcohol abuse, coagulopathies, seizures, cerebrospinal fluid shunts, metastases, frequent falls and the use of anticoagulant and antiplatelet therapies [37].

Nonetheless, the origin of CSDH is usually related with a previous head trauma in 60-80% of cases. The demographics of CSDH may explain the primary events that occur in this entity. On the one hand, brain atrophy, primarily present in elderly people, leads to a larger space between the surface of the brain and the dura matter. Furthermore, bridging veins (i.e., veins that go from the surface of the brain to the dural sinuses) are stretched as a result of the aforementioned brain atrophy, thus even a minor head trauma may produce a laceration of a bridging vein and, consequently, a bleeding in the subdural space. On the other hand, blood dyscrasias (due to medical therapy or a pathological condition) facilitate bleeding. Thus, CSDH is present in a specific population age group and it may be facilitated by the pathological conditions of the patients. This text provides a complete description of the CSDH pathophysiology.

The collection of fluid in the subdural space can produce brain hemisphere compression and, eventually, result in brain herniation. As there is a slow, progressive accumulation of fluid, CSDH can be clinically silent and the symptoms may appear insidiously, in the form of headache and varying degrees of neurological deficits. Psychiatric disturbances and epileptic seizures are also possible clinical manifestations.

CSDH is normally diagnosed by computed tomography (CT) scanning and CT scanning can be used to describe the different stages of CSDH. As suggested by Nomura et al. (1994) [21] the different forms of subdural collection may be of high density (acute subdural haematoma), isodensity (subacute subdural haematoma), low density, mixed density and layering type. The latter three are considered as the only forms of CSDH. Another noteworthy classification has been proposed by Nakaguchi et al. (2001) [19] who defined four groups of haematomas on the basis of CT scanning appearance: 1) homogeneous density type; 2) laminar type, defined as a subtype of homogeneous density, with a high density layer along the inner membrane; 3) layering or separated subtype, containing two components of different densities with a boundary lying between them; and 4) trabecular density type, in which a high -density septum between the inner and the outer membranes appeared against a low-density to isodense background. The difference in appearance could be related with different pathophsyological stages of the CSDH. This aspect will be further discussed in this chapter.

Although spontaneous resolution of CSDH has been described (mostly in small hematomas with no increase of intracranial pressure), surgical treatment is the main treatment option. There are different available surgical options: one/two burr hole/s with/without irrigation and with/without drainage; twist – drill craniostomy and craniotomy. There is no difference in outcome among the different surgical modalities with a recurrence rate of 4-26% and a serious associated morbidity [3]. Apart from surgery, there are also other medical therapies that have been described as useful in treating this condition. Among them, steroids are of specially interest. The importance of using medical therapies for these conditions lies in the possibility of avoiding a surgical procedure in patients where the surgery could be contraindicated and using steroids along with surgery to reduce the incidence of recurrence of the CSDH.

Bearing in mind the high incidence of CSDH and its particular pathophysiological features, the aim of this chapter is to describe the rationale of using steroids in the management of CSDH, its current role in the treatment of this condition and the potential of using this disease to investigate the effect of steroids in chronic local inflammatory processes.

Physiopathology of CSDH

The physiopathological mechanisms leading to a CSDH have been much discussed since Virchow's theory about "pachymeningitis" was published in 1857, which is now considered a classic, where the role of inflammation in the development of CSDH was first established. In fact, many authors considered CSDH as a chronic self-perpetuating local inflammatory process involving the dura matter, with elevation of pro-inflammatory factors, angiogenic factors and, finally, the formation of fibrinous tissue related with healing mechanisms.

As mentioned above, the origin of a CSDH is usually related with a head trauma and bleeding from a bridging vein. This trauma leads to a cleavage of the inner dural layer, creating a space that is normally considered as being virtual (the subdural space) [9, 27]. This subdural space can be created by an injury of the arachnoid membrane, as proposed by other authors [33]. In any case, the collection of blood and/or cephalic-spinal fluid (CSF) remains in direct contact with the inner dural border cell layer. This mesenchymal cell layer begins to proliferate and to form an inflammatory capsule or membrane around the blood clots or the CSF. This is called the external or outer membrane [16]. Different inflammatory cells (e.g., neutrophils, monocytes, macrophagues, fibroblasts, etc.), in this outer membrane form a type of granulation tissue. Furthermore, this membrane contains immature vessels, which have a great facility for bleeding. This last circumstance is clinically important, because when a CSDH is diagnosed signs of acute bleeding appear in the CT and this bleeding may be responsible for making the CSDH symptomatic. Therefore, a head trauma may lead to the development of a CSDH by a sequence of events consisting of local inflammation, angiogenesis and bleeding. These events are also associated with hypercoagulative activity, hyperfibrinolitic activity and increased vasopermeability, thus the local inflammation process is self-enhanced [10, 18, 34].

The role of inflammation in the physiopathology of CSDH has been reinforced by the determination of pro-inflammatory cytokines in the CSDH fluid. Some authors have demonstrated an elevation of IL-6, IL-8 and TNF-α (all of them pro-inflammatory cytokines) in the subdural fluid, while blood tests showed normal levels of these factors [32]. IL-6 is a pleiotrophic cytokine that influences immune and inflammatory responses and is one of the major physiological mediators of the acute phase reaction [15, 22]. Moreover, a direct pathogenic role of IL-6 in inflammatory angiogenesis and increase permeability has been inferred in other neurological pathological conditions [8]. On the other hand, IL-8 is considered the prototype of chemokines, i.e., factors presenting a chemotactic effect for migratory immune cells [2, 17]. IL-8 has also a very close relationship with the angiogenesis process [2, 17]. Both factors (IL-6 and IL-8) are secreted by fibroblasts and by endothelial and inflammtory cells infiltrating the outer membrane. In this sense, the local elevation of inflammatory factors shows that CSDH is a local inflammatory process, confined below the dura matter.

As mentioned before, both angiogenesis and vascular permeability play a critical role in the pathophysiology of the CSDH [11, 36]. The external neomembrane contains, among other inflammatory and repair related cells, fragile and leaky capillaries [26]. The formation of those capillaries is enhanced by the vascular endothelial growth factor (VEGF), a key inducer of angiogenesis and vascular permeability [14]. VEGF is upregulated in the CSDH fluid and it is also enhanced in neomembrane cells, as well as its receptor (VEGFR-1) [28, 31]. VEGF is not the only factor implicated in the angiogenesis process in CSDH. Other factors, such as the Placental Growth Factor (PlGF), also increase the VEGF response and appear in high concentrations in the CSDH fluid [14]. PIGF is usually induced under various pathological conditions associated with excessive and aberrant angiogenesis. Hypoxia-inducible factor (HIF)-1, a heterodimeric transcription factor induced among others by hypoxia, growth factors and oncogenes, positively regulates VEGF expression at the transcriptional level [38]. HIF-1 has also been shown to be over-expressed in the outer membrane of CSDH patients [20]. Therefore, different molecular pathways of the inflammatory response and angiogenesis are activated in CSDH. This process is self-regulated, showing different stages that differed in the degree of inflammation or angiogenic response and that may have clinical implications.

Although CSDH has been defined as a self-perpetuating local inflammation process, different stages are shown along the natural course of the disease, i.e., the inflammatory reaction does not always have the same intensity. This fact may even be evident in the CT scan. As explained above, CSDH appears in different forms in CT scanning (i.e., homogeneous, laminar, layering and trabecular). Each appearance could correspond to different stages in the inflammatory process, although this has not been clearly established. For example, higher concentrations of IL-6 and IL-8 were identified in layering CSDH and this was correlated with the risk of recurrence of the haematoma [8]. Other series have confirmed the higher rates of recurrence present in this type of CSDH [19, 21]. However, the lowest levels of these cytokines were measured in trabecular CSDH. Furthermore, the level of VEGF immunopositivity in the neomembrane is correlated with CSDH recurrence [14].

The layering appearance in CT scanning of a CSDH seems to indicate a more active inflammatory reaction and a more intense angiogenic process and, consequently, a higher risk of recurrence. This fact is crucial, because the use of anti-inflammatory/anti-angiogenic therapies should be adapted to the degree of inflammation/angiogenesis existing in each case. In fact, the degree of inflammation may be the main prognostic factor for recurrence.

Furthermore, layering haematomas show shorter median intervals between the trauma and the onset of symptoms, while trabecular CSDH has the longest intervals [8, 19]. This seems to show that trabecular haematoma might be the most chronic stage of a CSDH where the inflammatory process is less intense and there is a prevalence of fibrotic phenomena within the neomembrane. Thus, recurrence rates are lower (the tendency to bleed is also lower) and the anti-inflammatory therapy would be much less effective.

Therefore, inflammation and angiogenesis are the two key factors in the pathophysiology of the CSDH. Therefore, therapies that modify or modulate these responses should be considered in the treatment of CSDH, mostly when they are very intense since the recurrence risk is very high or when the surgical treatment may be associated with important co-morbidities. Nevertheless, the choice of these therapies should consider the stage of the natural course of the CSDH because the intensity of the inflammatory response and the angiogenic process vary along the course of the CSDH.

Rationale of the Use of Steroids in CSDH Treatment

The basis to use steroids in CSDH is their anti-inflammatory capacity. As it has been explained previously, CSDH can be considered as a chronic local inflammatory disease. Steroids are supposed to inhibit the production of pro-inflammatory cytokines and to induce of anti-inflammatory cytokines production, thus the most important pharmacologic property of steroids is their immunosupresive effect [23]. Further, they promote phagocytosis of apoptotic leukocytes, thus steroids are also considered as an agent involved in the resolution phase of inflammation [29].

The action of glucocorticoids is complex and depends on the induction of anti-inflammatory regulatory proteins as well as inhibition of signalling pathways such as NF-kβ[1] and AP-1[2]. Other important glucocorticoid induced protein is AnxA1. This protein has shown to have anti-inflammatory and pro-resolving properties in various animal models of inflammation and in physiological conditions. AnxA1 mediates cell apoptosis and efferocytosis and it has been shown to be induced by dexamethasone [35].

Talking specifically about CSDH, dexamethasone has also demonstrated to produce an inhibition of neomembrane formation in murine models [5]. Although the inhibition of all the above mentioned signalling pathways has not been specifically demonstrated when using dexamethasone in CSDH, they are probably implicated in the formation of neomembranes in CSDH, and therefore the use of steroids would block one of the main pathophysiological steps in the evolution of CSDH (i.e., neomembrane formation through inflammatory pathways) and a resolution of the disease could be achieved with their use.

Furthermore, steroids can also be useful in reducing neo-angiogenesis. This antiangiogenic effect is thought to be a consequence of multiple anti-inflammatory properties including, among others, inhibition of cell chemotaxis and modulation of the proteolytic activities of vascular endothelial cells that precedes the budding of new vessels [13]. In this sense, steroids have been shown as a negative modulator of the expression of VEGF [24]. Importantly, the potency of the antiangiogenic effect of the steroids is independent of their relative glucocorticoid and mineralocorticoid activity [13].

Apart from the anti-inflammatory and anti-angiogenic action of steroids, they also induce the secretion of the inhibitor of plasminogen, a substance that reduces rebleeding-lysis cycle of the clot [6].

Because of the aforementioned, steroids seem to be an appropriate treatment option in the management of CSDH because they can interfere in many of the pathological pathways that this entity presents. However, the evaluation of clinical studies is needed to identify what is really the role of steroids in the management of CSDH.

---

[1] NF Kβ *(nuclear factor kappa-light-chain-enhancer of activated B cells)* is a protein complex that controls ADN transcription. It is implicated in the cellular response mediated by cytokines (among other stimulus).

[2] AP-1 *(activator protein 1)* is another transcription factor composed by different proteins. Its production is stimulated by pro-inflammatory cytokines and it is implicated in numerous cellular processes like proliferation, differentiation and apoptosis.

Clinical Evidence for Using Steroids in CSDH Management

The use of steroids in CSDH can be justified by the pathophysiology of the CSDH and the anti-inflammatory properties of steroids. Both aspects have been previously extensively discussed. However, the use of this sort of therapy in the clinical scenario is far from being standardized. Attention has been brought to this situation by surveys from the United Kingdom, Ireland, Canada and France, where approximately 50% of neurologists and neurosurgeons never use corticosteroids [4, 12, 25]. The lack of randomized clinical trials and the fear of steroid-associated-adverse effects are probably the reasons why steroids are not more widely used in the management of CSDH.

Bearing in mind that surgery is the standard treatment for CSDH and that good results have been widely reported with different surgical procedures, the use of steroids in CSDH has been reserved for three scenarios: minimally symptomatic or asymptomatic patients; for minimum radiological and clinical recurrences after surgical drainage; and in surgically contraindicated patients. However, despite the good results obtained by surgery, complications may occur and some of them may be potentially severe or fatal. Therefore, there are two possible indications for using corticosteroids in CSDH. On the one hand, using steroids as a single treatment without any sort of surgical procedure; and on the other hand, using steroids in the peri-operative time, as an adjuvant therapy of surgery.

Steroids As the Primary Treatment

There are a few studies that have focused on the role of steroids as a single treatment for CSDH. Most of them consist of case reports or small series of patients where the use of steroids solved the CSDH without needing surgery. More recent reports have described the same results with a larger number of patients. In this sense, in 2005, Sun et al. concluded that corticosteroids (particularly dexamethasone) could be a medical alternative for selected symptomatic CSDH patients who are not suitable for surgical intervention (elderly patients with medical co-morbidity or who refuse surgical treatment), although there was not a valid comparison in this study of the effectiveness of medical and surgical treatment alternatives in this population of CSDH patients [30].

Four years later, in the study of Delgado-Lópex PD et al. (2009), 122 CSDH were retrospectively reviewed. Those cases were treated following an internal protocol where dexamethasone was administered in patients with good neurological status while worse clinical cases were directly treated with surgery. The group of patients treated with steroids was re-evaluated 48-72 hours after administration and the main variable of the study was the outcome of the patients. Ninety-six percent of patients assigned to treatment with dexamethasone presented a favourable outcome. However, among those who were initially treated with dexamethasone, 21.8% eventually required surgical treatment. The study of Delgado-Lópex PD et al. (2009) shows that many patients with CSDH can be treated with steroids with good results, but no comparison with surgical treatment could be made because of methodological limitations [6].

However, a recent meta-analysis concluded that using steroids as the main management plan did not result in a reduction of mortality or morbidity, with improvement in neither cure nor recurrence rates [1]. Nevertheless, the authors themselves suggest interpreting those results cautiously, as data were scarce and abstracted from a small number of observational studies [1].

Therefore, although properly-designed randomized clinical trials must be performed to collect better evidence, using steroids as the primary therapy for CSDH seems to be a plausible option, mostly in patients with minor neurological symptoms and/or patients who have high surgical-related risks.

Steroids As Adjuvant Treatment

Steroids have also been used concomitantly with surgery in many observational studies. As is the case of steroids as primary treatment, using steroids as an adjuvant treatment for CSDH has not been investigated by randomized clinical trials, thus there is no solid evidence of their effectiveness as an adjuvant treatment. In any case, the rationale for using steroids in combination with surgery is the same for using steroids on their own. The anti-inflammatory and anti-angiogenic properties of steroids act in this local inflammation process where surgical evacuation (any sort of the described surgical procedures) has rapidly or progressively eliminated the inflammatory factors perpetuating the CSDH.

Bearing this in mind, Berhauser et al. (2012) made a comparison between patients treated with burr hole craniostomy alone or combined with peri-operative dexamethasone. The time exposed to steroids was also measured and the authors found that the longer exposition of steroids prior to surgery was associated with lower rates of recurrences. Furthermore, no association between the use of peri-operative steroids and post-operative complications was shown [3]. On the contrary, a meta-analysis of 17 pooled cohorts does not support the case for a favourable outcome for recommending the use of steroids as an adjuvant therapy. In fact, higher morbidities were associated with the use of corticosteroids combined with surgical management [1]. However, these results have to be considered carefully due to the lack of randomized clinical trials and the heterogeneity of the studies included in the meta-analysis. At present, the DRESH study (a clinical trial which is designed to answer the question whether the use of dexamethasone reduces the recurrences rates) is ongoing [7]. The DRESH study will probably provide more evidence of the possible benefits of using steroids as an adjuvant treatment of surgery in the management of CSDH. Furthermore, everyone knows that the use of steroids is associated with a number of morbidities. When they are used in CSDH, the most reported complications have been hypertension and hyperglycaemia which are difficult to control in diabetic patients. Chronic-intake related complications are not normally present, because the use of steroids for this condition is for a limited period of time. Therefore, considering the successful results of previous observational studies and the weakness of the meta-analysis of Almenawer et al. (2014), adjuvant treatment of CSDH with steroids may be recommended to prevent recurrences, although special care should be taken with diabetic patients.

Future Perspectives

The role of steroids in CSDH has not been definitely determined yet. Although its pathophysiology provides a good rationale for using it as a part of CSDH treatment, there are still many inconsistencies regarding its clinical application. These inconsistencies could be related with the use of steroids in cases of CSDH with a low intense inflammatory reaction. We can suggest that this treatment may be more useful in patients who show a higher intensity inflammatory reaction, which is related with higher rates of recurrences. The use of steroids in such patients could reduce hematoma recurrence. In this sense, a layering appearance in CT scanning of a CSDH seems to indicate a more active inflammatory reaction and a more intense angiogenic process and, consequently, a higher risk of recurrence (as mentioned above). Bearing this in mind, depending on the image of the CSDH in the CT scanning, the clinician might be able to determine the risk of recurrence and to decide whether an anti-inflammatory therapy (like steroids) is suitable for this case or not.

In any case, there is a need for randomized clinical trials which define whether steroids have a role in the management of CSDH better. Study protocols should consider the degree of inflammation (based on CT scanning appearance and cytokines measurements) as a possible confused variable. Apart from this, clinical series should appropriately differentiate among pre-surgical use of steroids, post-surgical use of steroids, peri-operative use of steroids (pre-surgical and post-surgical) and the use of steroids as the only treatment. This would make it possible to identify the different efficacies of steroid treatment with respect to when they are administered. This issue, to best of our knowledge, has still not been studied. On the other hand, bearing in mind that CSDH is a chronic local inflammatory process, CSDH could be a good scenario to further investigate the pathophysiological pathways that are involved in this sort of inflammatory condition. Greater knowledge of these pathways could enable a deeper study of the molecular mechanisms used by steroids and the identification of targets for other anti-inflammatory drugs.

Conclusion

CSDH is a common neurosurgical disease whose primarily treatment is surgery. However, in some patients, steroids are used as the only treatment or as an adjuvant treatment for surgery. The use of steroids seems to be associated with lower levels of recurrences in surgically treated hematomas and they have proved to be effective when used on their own in the treatment of some patients. The rationale for using steroids in the management of CSDH is based on their inflammatory-related pathophysiology, where higher levels of various pro-inflammatory and neo-angiogenesis molecular factors have been locally determined. In spite of this, the use of steroids is not well standardized and more clinical evidence is needed to support and clarify their role in CSDH management.

References

[1] Almenawer, S. A., Farrokhyar, F., Hong, C., Alhazzani, W., Manoranjan, B., Yarascavitch, B., Arjmand, P., Baronia, B., Reddy, K., Murty, N.,

Singh, S. (2014) Chronic subdural hematoma management: a systematic review and meta-analysis of 34,829 patients. *Annals of Surgery*, *259*, 449–57.

[2] Baggiolini, M., Dewald, B., & Moser, B. (1994) Interleukin-8 and related chemotactic cytokines--CXC and CC chemokines. *Advances in Immunology*, *55*, 97–179.

[3] Berghauser Pont, L. M. E., Dirven, C. M. F., Dippel, D. W. J., Verweij, B. H., & Dammers, R. (2012) The role of corticosteroids in the management of chronic subdural hematoma: a systematic review. *European Journal of Neurology: The Official Journal of the European Federation of Neurological Societies*, *19*, 1397–403.

[4] Cenic, A., Bhandari, M., & Reddy, K. (2005) Management of chronic subdural hematoma: a national survey and literature review. *The Canadian Journal of Neurological Sciences. Le Journal Canadien Des Sciences Neurologiques*, *32*, 501–6.

[5] D'Abbondanza, J. A., & Loch Macdonald, R. (2014) Experimental models of chronic subdural hematoma. *Neurological Research*, *36*, 176–88.

[6] Delgado-López, P. D., Martín-Velasco, V., Castilla-Díez, J. M., Rodríguez-Salazar, A., Galacho-Harriero, A. M., & Fernández-Arconada, O. (2009) Dexamethasone treatment in chronic subdural haematoma. *Neurocirugía (Asturias, Spain)*, *20*, 346–59.

[7] Emich, S., Richling, B., McCoy, M. R., Al-Schameri, R. A., Ling, F., Sun, L., Wang, Y., Hitzl, W. (2014) The efficacy of dexamethasone on reduction in the reoperation rate of chronic subdural hematoma--the DRESH study: straightforward study protocol for a randomized controlled trial. *Trials*, *15*, 6.

[8] Frati, A., Salvati, M., Mainiero, F., Ippoliti, F., Rocchi, G., Raco, A., Caroli, E., Cantore, G., Delfini, R. (2004) Inflammation markers and risk factors for recurrence in 35 patients with a posttraumatic chronic subdural hematoma: a prospective study. *Journal of Neurosurgery*, *100*, 24–32.

[9] Friede, R. L., & Schachenmayr, W. (1978) The origin ofsubdural neomembranes. II. Fine structural of neomembranes. *The American Journal of Pathology*, *92*, 69–84.

[10] Fujisawa, H., Nomura, S., Tsuchida, E., & Ito, H. (1998) Serum protein exudation in chronic subdural haematomas: a mechanism for haematoma enlargement? *Acta Neurochirurgica*, *140*, 161–5; discussion 165–6.

[11] Funai, M., Osuka, K., Usuda, N., Atsuzawa, K., Inukai, T., Yasuda, M., Watanabe, Y., Takayasu, M. (2011) Activation of PI3 kinase/Akt

signaling in chronic subdural hematoma outer membranes. *Journal of Neurotrauma*, *28*, 1127–31.

[12] Guénot, M. (2001) [Chronic subdural hematoma. Introduction and results of a survey by the French Society of Neurosurgery]. *Neuro-Chirurgie*, *47*, 459–60.

[13] Gupta, D., & Illingworth, C. (2011) Treatments for corneal neovascularization: a review. *Cornea*, *30*, 927–38.

[14] Kalamatianos, T., Stavrinou, L. C., Koutsarnakis, C., Psachoulia, C., Sakas, D. E., & Stranjalis, G. (2013) PlGF and sVEGFR-1 in chronic subdural hematoma: implications for hematoma development. *Journal of Neurosurgery*, *118*, 353–7.

[15] Kishimoto, T. (1989) The biology of interleukin-6. *Blood*, *74*, 1–10.

[16] Kwon, T. H., Park, Y. K., Lim, D. J., Cho, T. H., Chung, Y. G., Chung, H. S., & Suh, J. K. (2000) Chronic subdural hematoma: evaluation of the clinical significance of postoperative drainage volume. *Journal of Neurosurgery*, *93*, 796–9.

[17] Matsushima, K., & Oppenheim, J. J. (1989) Interleukin 8 and MCAF: novel inflammatory cytokines inducible by IL 1 and TNF. *Cytokine*, *1*, 2–13.

[18] Moskala, M., Goscinski, I., Kaluza, J., Polak, J., Krupa, M., Adamek, D., Pitynski, K., Miodonski, A. J. (2007) Morphological aspects of the traumatic chronic subdural hematoma capsule: SEM studies. *Microscopy and Microanalysis : The Official Journal of Microscopy Society of America, Microbeam Analysis Society, Microscopical Society of Canada*, *13*, 211–9.

[19] Nakaguchi, H., Tanishima, T., & Yoshimasu, N. (2001) Factors in the natural history of chronic subdural hematomas that influence their postoperative recurrence. *Journal of Neurosurgery*, *95*, 256–62.

[20] Nanko, N., Tanikawa, M., Mase, M., Fujita, M., Tateyama, H., Miyati, T., & Yamada, K. (2009) Involvement of hypoxia-inducible factor-1alpha and vascular endothelial growth factor in the mechanism of development of chronic subdural hematoma. *Neurologia Medico-Chirurgica*, *49*, 379–85.

[21] Nomura, S., Kashiwagi, S., Fujisawa, H., Ito, H., & Nakamura, K. (1994) Characterization of local hyperfibrinolysis in chronic subdural hematomas by SDS-PAGE and immunoblot. *Journal of Neurosurgery*, *81*, 910–3.

[22] Patterson, P. H. (1992) The emerging neuropoietic cytokine family: first CDF/LIF, CNTF and IL-6; next ONC, MGF, GCSF? *Current Opinion in Neurobiology*, *2*, 94–7.

[23] Saffar, A. S., Ashdown, H., & Gounni, A. S. (2011) The molecular mechanisms of glucocorticoids-mediated neutrophil survival. *Current Drug Targets*, *12*, 556–62.

[24] Santarius, T., & Hutchinson, P. J. (2004) Chronic subdural haematoma: time to rationalize treatment? *British Journal of Neurosurgery*, *18*, 328–32.

[25] Santarius, T., Lawton, R., Kirkpatrick, P. J., & Hutchinson, P. J. (2008) The management of primary chronic subdural haematoma: a questionnaire survey of practice in the United Kingdom and the Republic of Ireland. *British Journal of Neurosurgery*, *22*, 529–34.

[26] Sato, S., & Suzuki, J. (1975) Ultrastructural observations of the capsule of chronic subdural hematoma in various clinical stages. *Journal of Neurosurgery*, *43*, 569–78.

[27] Schachenmayr, W., & Friede, R. L. (1978) The origin of subdural neomembranes. I. Fine structure of the dura-arachnoid interface in man. *The American Journal of Pathology*, *92*, 53–68.

[28] Shono, T., Inamura, T., Morioka, T., Matsumoto, K., Suzuki, S. O., Ikezaki, K., Iwaki, T., Fukui, M. (2001) Vascular endothelial growth factor in chronic subdural haematomas. *Journal of Clinical Neuroscience : Official Journal of the Neurosurgical Society of Australasia*, *8*, 411–5.

[29] Sousa, L. P., Alessandri, A. L., Pinho, V., & Teixeira, M. M. (2013) Pharmacological strategies to resolve acute inflammation. *Current Opinion in Pharmacology*, *13*, 625–31.

[30] Sun, T. F. D., Boet, R., & Poon, W. S. (2005) Non-surgical primary treatment of chronic subdural haematoma: Preliminary results of using dexamethasone. *British Journal of Neurosurgery*, *19*, 327–33.

[31] Suzuki, K., Takano, S., Nose, T., Doi, M., & Ohashi, N. (1999) Increased concentration of vascular endothelial growth factor (VEGF) in chronic subdural hematoma. *The Journal of Trauma*, *46*, 532–3.

[32] Suzuki, M., Endo, S., Inada, K., Kudo, A., Kitakami, A., Kuroda, K., & Ogawa, A. (1998) Inflammatory cytokines locally elevated in chronic subdural haematoma. *Acta Neurochirurgica*, *140*, 51–5.

[33] Tanaka, Y., & Ohno, K. (2013) Chronic subdural hematoma - an up-to-date concept. *Journal of Medical and Dental Sciences*, *60*, 55–61.

[34] Tokmak, M., Iplikcioglu, A. C., Bek, S., Gökduman, C. A., & Erdal, M. (2007) The role of exudation in chronic subdural hematomas. *Journal of Neurosurgery*, *107*, 290–5.

[35] Vago, J. P., Nogueira, C. R. C., Tavares, L. P., Soriani, F. M., Lopes, F., Russo, R. C., Pinho, V., Teixeira, M. M., Sousa, L. P. (2012) Annexin A1 modulates natural and glucocorticoid-induced resolution of inflammation by enhancing neutrophil apoptosis. *Journal of Leukocyte Biology*, *92*, 249–58.

[36] Weigel, R., Hohenstein, A., Schlickum, L., Weiss, C., & Schilling, L. (2007) Angiotensin converting enzyme inhibition for arterial hypertension reduces the risk of recurrence in patients with chronic subdural hematoma possibly by an antiangiogenic mechanism. *Neurosurgery*, *61*, 788–92; discussion 792–3.

[37] Zarkou, S., Aguilar, M. I., Patel, N. P., Wellik, K. E., Wingerchuk, D. M., & Demaerschalk, B. M. (2009) The role of corticosteroids in the management of chronic subdural hematomas: a critically appraised topic. *The Neurologist*, *15*, 299–302.

[38] Ziello, J. E., Jovin, I. S., & Huang, Y. (2007) Hypoxia-Inducible Factor (HIF)-1 regulatory pathway and its potential for therapeutic intervention in malignancy and ischemia. *The Yale Journal of Biology and Medicine*, *80*, 51–60.

In: Corticosteroids and Steroid Therapy
Editor: Carmen Adkins
ISBN: 978-1-63482-308-1
© 2015 Nova Science Publishers, Inc.

Chapter 4

Early Diagnosis and Preventive Strategy of Corticosteroid Induced Osteonecrosis in Systemic Autoimmune Diseases

Syuichi Koarada*[*], *Yukiko Tokuda, Yukihide Ono, Yuri Sadanaga, Satoko Tashiro, Rie Suematsu, Nobuyuki Ono, Akihide Ohta and Yoshifumi Tada
Division of Rheumatology, Faculty of Medicine,
Saga University, Saga, Japan

Abstract

Osteonecrosis of femoral head (ONF) is one of the serious adverse events in the patients with systemic lupus erythematosus (SLE) associated with corticosteroid therapy. We have reported a multicenter prospective study of prevention of ONF in SLE patients on high doses of corticosteroids using anticoagulant of warfarin. In the diagnosis of ONF, plain radiography and magnetic resonance imaging (MRI) are important. Especially, in early stage of ONF, although the plain radiograph is still normal, evident changes can be seen in MRI. The treatment of ONF

[*]Corresponding author: Division of Rheumatology, Faculty of Medicine, Saga University, 5-1-1 Nabeshima, Saga 849-8501, Japan; Tel: +81-(952)34-2367; Fax: +81-(952)34-2017.

remains controversial. Anticoagulants may be useful to prevent ONF. Therefore, early diagnosis and prevention of ONF are critical issues especially in SLE patients. In this chapter, we present the radiological images illustrating osteonecrosis in patients with autoimmune diseases including SLE, and review the strategy to prevent ONF induced by corticosteroids.

Introduction

Osteonecrosis is characterized by bone death as a result of a compromised artery supply. Osteonecrosis is also known as avascular necrosis (AVN), aseptic necrosis, subchondral avascular necrosis, or ischemic necrosis of bone. Vascular interruption of the blood supply to the bone is followed by reactive hyperemia and bone necrosis leading to subchondral fractures. It results in flattening of the bone surface and subsequent degenerative changes of the bone and adjacent various structures [1, 2].

Systemic lupus erythematosus (SLE) is one of the prototypical systemic autoimmune diseases characterized by great heterogeneity involving multiple organs and apparatuses. [3]. Osteonecrosis of femoral head (ONF) is a serious complication of SLE especially associated with corticosteroid therapy [4].There have been many reports about the prevalence of ONF in SLE patients ranging 4% to 40% [5-12]. However, the approximate average is 10% as a whole by the diagnosis using clinical symptoms and traditional plain radiography [8].

Magnetic resonance imaging (MRI) dramatically changed the epidemiology of osteonecrosis in SLE patients and the prevalence of ONF using MRI appears to be considerably higher than the diagnosis using conventional technology. MRI provides the diagnosis of osteonecrosis more sensitively. Previously, we have described the early development of corticosteroid induced ONF in SLE patients using MRI [4]. ONF occurs very early in one-third of SLE patients with high doses of corticosteroids, especially corticosteroid pulse therapy.

There have been a few studies of the strategy of prevention and therapy for osteonecrosis in SLE patients induced by high doses of corticosteroid. Anticoagulants, warfarin or enoxaparin, may be potential drugs to prevent ONF [4]. Both early diagnosis and prevention of ONF are critical issues in SLE patients treated with corticosteroids. We present the fundamental radiological images illustrating osteonecrosis and review the strategy to prevent ONF induced by corticosteroids in patient with autoimmune diseases including SLE.

Osteonecrosis in SLE Patients Treated with Corticosteroids

Osteonecrosis may begin by interruption of blood supply causing bone ischemia in various systemic diseases and other conditions. Abnormalities in lipid metabolism, bone homeostasis, regulation of apoptosis, coagulopathies, and oxidative stress may play roles in the pathogenesis of osteonecrosis. However, the final common pathway of osteonecrosis is disruption of blood supply to a segment of bone.

SLE patients are at high risk of developing osteonecrosis because of both the disease for itself and corticosteroid therapy. More importantly, osteonecrosis develops in a relatively short time after the starting of high doses of corticosteroids [13-15].

The duration between the initiation of corticosteroids and the development of osteonecrosis ranges from 1 to 16 months [13, 16, 17].

However, osteonecrosis does not develop in SLE patients if the dose of corticosteroid is maintained low. New cases of osteonecrosis were not observed afterward [13]. Enlargement of osteonecrosis occurs only after increasing corticosteroid dosage [18]. A study of 10-year minimum follow-up with MRI showed spontaneous repair of asymptomatic osteonecrosis in SLE patients [18]. At final follow-up, half of the lesions (49%) demonstrated spontaneous repair in the necrotic area. Complete regression of osteonecrosis was observed in 9% of the cases. Although medium or large area of osteonecrosis progresses and finally results in the collapse, a small fragment of osteonecrosis may be asymptomatic without progress.

The corticosteroid dose was not significantly associated with osteonecrosis [13]. Because many other factors affect the development of osteonecrosis, analysis of dose-response risk for an isolated association is difficult. However, corticosteroid-induced osteonecrosis may be dependent on dosage, the long-acting steroids, and parenteral usage.

Signs and Symptoms

The primary symptom of osteonecrosis is pain. The severity of clinical signs and symptoms in osteonecrosis depends on the anatomic region and size of osteonecrosis. Osteonecrosis confined to medullary bone, bone infarct, can be asymptomatic. The onset of pain due to osteonecrosis is insidious or sudden. In general, the pain is a gradual onset characterized by mild vague and slow incremental progression. The pain may worsen with use of the joint by weight-bearing and ambulation. The pain can be persistent even at rest in the advanced cases of osteonecrosis. The pain from acute osteonecrosis in deep joints can be strong. Although joint range of motion is preserved in early stages, limitation of range of motion is usually a progressive and late symptom.

Radiographs of Osteonecrosis

Ficat Staging System

Although various classifications of the staging in osteonecrosis have been proposed, the Ficat Staging System (stage I to stage IV) for the radiograph is widely used (table 1) [19].

Stage I: Normal radiographs. Conventional radiographs are normal. The patient is usually asymptomatic or may have minimal pain.

Stage II: Sclerotic or cystic lesions. In radiographs, osteosclerotic or cystic lesions without subchondral fracture are found.

Table 1. Ficat Staging System

Stage I	Normal radiographs
Stage II	Sclerotic or cystic lesions
Stage III	Subchondral collapse
Stage IV	Osteoarthritis with articular collapse

Stage III: Subchondral collapse. Radiographs show the typical “crescent sign” due to collapse of a necrotic segment of subchondral trabecular bone. However, joint space remains intact.

Stage IV: Osteoarthritis with articular collapse. Radiographs show terminal stage of osteoarthritic changes with collapse.

Steinberger Staging System

The classification system of Ficat Staging System has been refined by other groups. Steinberg M. et al. proposed another classification system (stage I to stage VI) that modified the Ficat system by including bone scintigraphy and MRI, and volumetric assessment of the femoral head [20].

Table 2. Steinberger Staging System

Stage I.	Normal radiograph; abnormal bone scan and/or MRI
A	Mild (<15% of head affected)
B	Moderate (15-30% of head affected)
C	Severe (>30% of head affected)
Stage II	Lucent and sclerotic changes in the femoral head
A	Mild (<15% of head affected)
B	Moderate (15-30% of head affected)
C	Severe (>30% of head affected)
Stage III.	Subchondral collapse without flattening
A	Mild (<15% of head affected)
B	Moderate (15-30% of head affected)
C	Severe (>30% of head affected)
Stage IV.	Flattening of the femoral head
A	Mild (<15% of head affected)
B	Moderate (15-30% of head affected)
C	Severe (>30% of head affected)
Stage V.	Joint narrowing and/or acetabular changes
A	Mild
B	Moderate
C	Severe
Stage VI.	Advanced degenerative changes

The Association Research Circulation Osseus (ARCO), The International Classification of Osteonecrosis of the Femoral Head

The International Classification of Osteonecrosis of the Femoral Head, the Association of Research Circulation Osseous (ARCO) modification, adopts an osteonecrosis staging system and adds a Stage 0 (Normally no pain, all imaging normal, and bone biopsy results consistent with osteonecrosis) at high risk for development of ONF [21].

Table 3. Association Research Circulation Osseus International Classification of Osteonecrosis of the Femoral Head

Stage	Clinical	Findings
0	Normally no pain	All imaging normal; bone biopsy: plasmostasis and marrow necrosis
I	Normally no pain	Positive findings on MRI and bone scan; radiograph normal
II	May have pain	Positive findings on MRI and bone scan; radiograph normal
III	Pain	Crescent sign and/or flattening of articular surface of femoral head on radiograph
IV	Pain	Radiographic evidence of osteoarthritis: joint space narrowing, acetabular changes and joint destruction

Imaging Studies of Osteonecrosis in SLE

The diagnosis of osteonecrosis is confirmed by imaging studies including plain radiographs, bone scintigraphy, computed tomography (CT), and MRI. Plain radiographs have been used for the diagnosis and evaluation of osteonecrosis. Anteroposterior (AP) and frog leg views provide an evaluation of the morphology and quality of the femoral head. These radiographs should be performed for the screening of osteonecrosis of the hip [22].

A. Alignment

Stage 0 - II (ARCO staging): In early stages, alignment in conventional radiographs is normal. Stage III: There is flattering of articular surface of femoral head due to subchondral collapse. Stage IV: With the development of the disease, radiographs show terminal stage of secondary osteoarthritic changes with collapse (Figure 1). Total destruction of the involved structure such as the femoral head may be observed.

B. Bone Density

In radiographs of early stage (stage II), osteosclerotic or cystic lesions are found. The earliest radiographic finding may be smudgy density of the femoral heads (Figure 2).

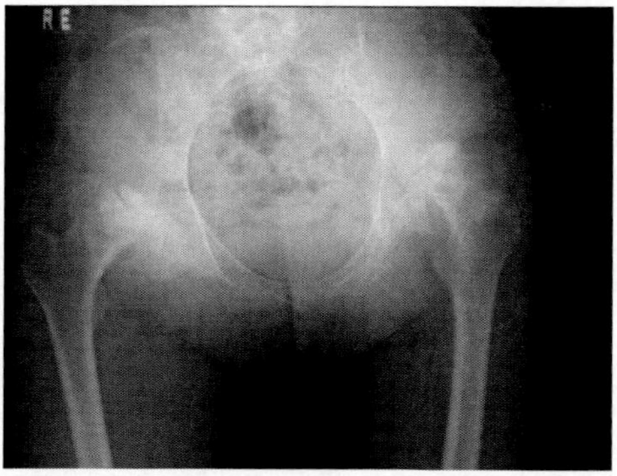

Figure. 1. The AP view of the pelvis in a patient with SLE demonstrates increased density of the both femoral heads. The radiograph shows destruction of the normal spherical shape of femoral heads and flattering of the superior aspect due to osteonecrosis (stage IV). Secondary osteoarthritis with joint space narrowing is also observed.

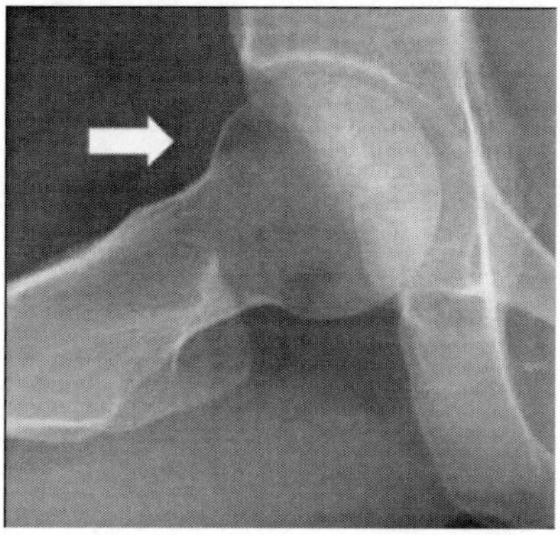

Figure 2. Plain radiograph of the right hip in a patient with SLE demonstrates very early osteonecrosis of the femoral head (Steinberg stage II). Sclerotic areas with vague increased smudgy density and radiolucent parts are observed in the right femoral head. In osteonecrosis, the first radiographic change is smudging of the trabecular pattern near the articular surface of the humeral head.

In stage III, the area of ischemic necrosis appears dense in comparison with the remaining viable portion or reparative bone. The osteonecrotic area is surrounded by a reactive margin of variable but focally increased density. The typical radiographic sign of osteonecrosis is crescent sign. It is the presence of a radiolucent crescent shaped rim along the contour of the femoral head (Figure 3). At stage III, the changes are already irreversible.

Increased density of the femoral heads as sclerotic changes is typical in advanced osteonecrosis (stage IV). The appearance is secondary to compression of bone trabeculae, calcification, and repair of the necrotic area by deposition of new bone.

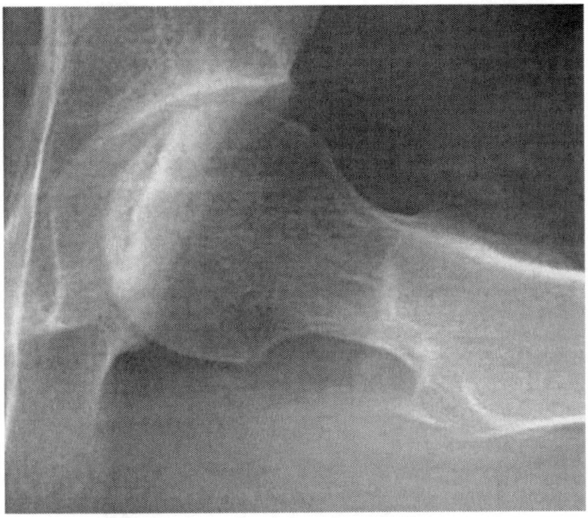

Figure 3. The flog leg radiograph of the left femoral head in an SLE patient with osteonecrosis shows the radiolucent "crescent sign" (stage III).

Bone Infarct

Radiographs of bone infarcts show a lytic appearance with serpiginous and well-defined geographic sclerotic borders (Figure 4 and 5).

C. Cartlidge

In stage 0 to III, joint space remains intact (Figure 6). However, collapse adjacent to articular cartilage leads to secondary osteoarthritis (stage IV). Additionally, joint effusion is found in approximately 60% of cases [23].

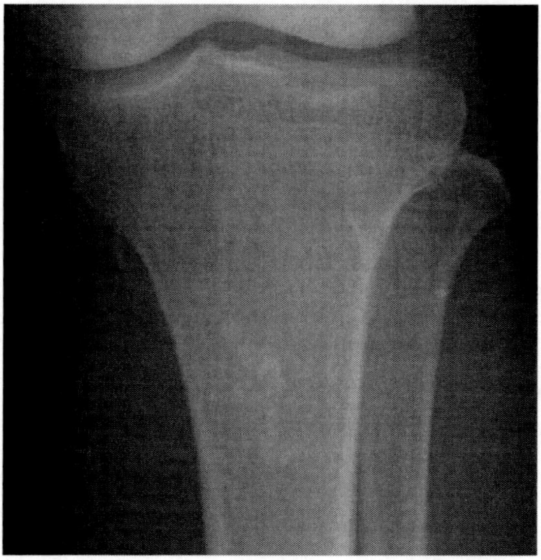

Figure 4. The AP view of the knee in a patient with SLE shows the area of sclerotic curvilinear densities due to medullary infarct of tibia. The finding indicates a possible bone infarct.

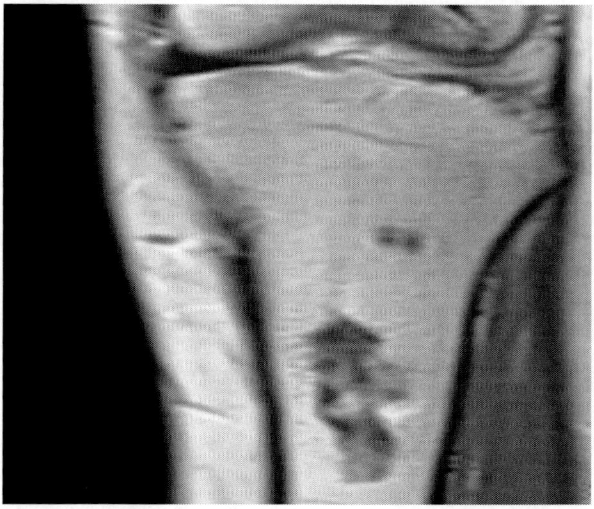

Figure 5. The bone infarct of tibia is confirmed with coronal T1-weighted MRI of the knee.

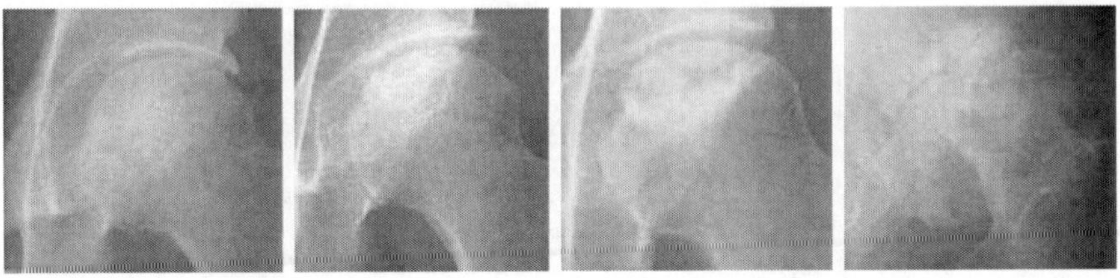

Figure 6. In stage III, increased density is observed in the left femoral head (Stage I to IV). However, joint space remains intact until terminal stage. The radiograph of stage IV demonstrates the femoral head collapse.

D. Distribution

Osteonecrosis in SLE patients affects the femoral heads as the most frequent site. However, osteonecrosis can be bilateral and multifocal in SLE. Osteonecrosis develops in the femoral condyles, the tibial plateaus, distal ends of the tibia, the tali, and the humeral heads [24-27]. Also, Osteonecrosis occurs in the lunates, the scaphoids, the metacarpal heads and metatarsal heads on occasion [11, 28, 29].

In some cases, bone infarct can be observed. Bone infarct occurs within the diaphysis or metadiaphysis of the bone. The characteristics of bone infarcts share many of the predisposing factors of osteonecrosis. Because multifocal involvement is common in SLE patients, other lesions should be screened (Figure 7-10).

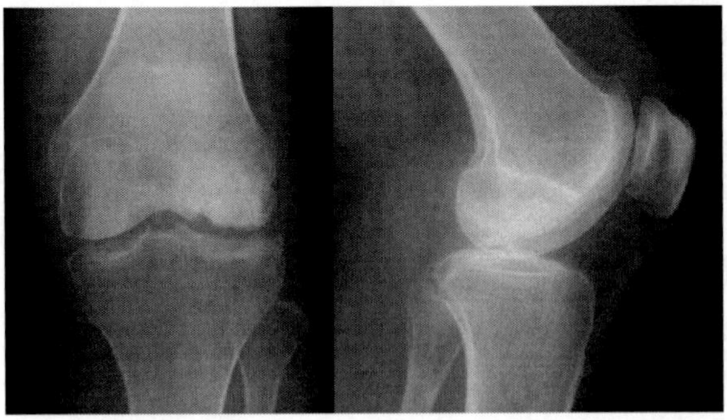

Figure 7. The AP and lateral views of the left knee in an SLE patient show osteonecrosis of the lateral femoral condyle. Although the lateral condyle is deformed, the joint space is still maintained.

E. External Bone: Soft Tissue

There is not characteristic finding of osteonecrosis in soft tissues.

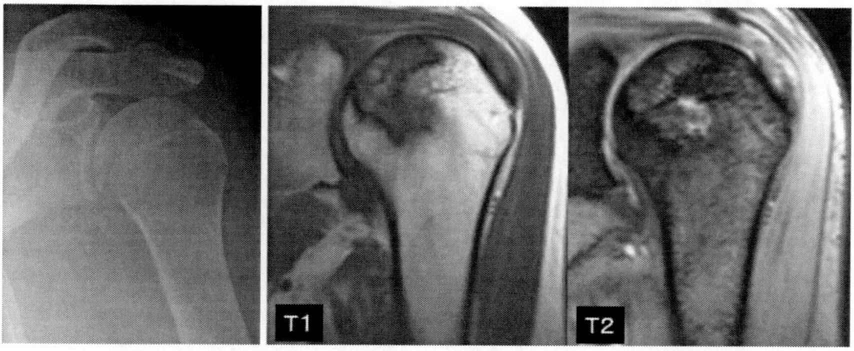

Figure 8. The plain radiograph of the left shoulder shows no abnormal findings. Coronal T1- and T2-weighted images of the left shoulder show osteonecrosis.

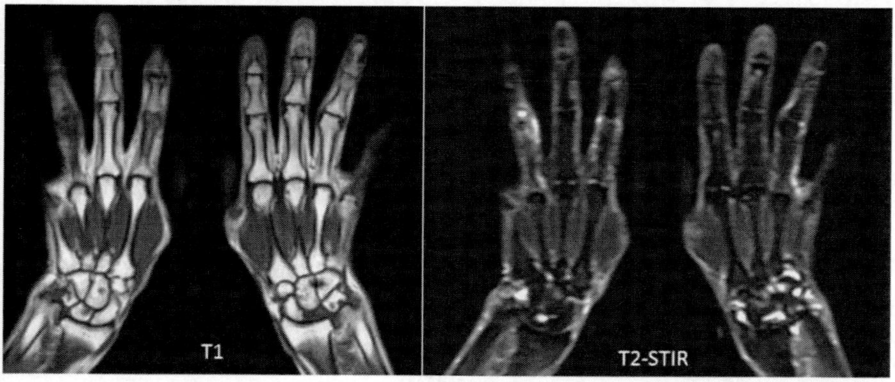

Figure 9. Osteonecrosis of the lunate. The T1-weighted coronal MR image shows low intensity of the lunate. The MRI is more sensitive than plain film to detect early change of osteonecrosis.

F. Further Examination

Blood Tests

In the blood tests, high doses of corticosteroids may cause elevation of the levels of total cholesterol, albumin, and leukocyte count in most of the SLE patients. Elevation of total cholesterol seems to be associated with the pathogenesis of ONF. [30]

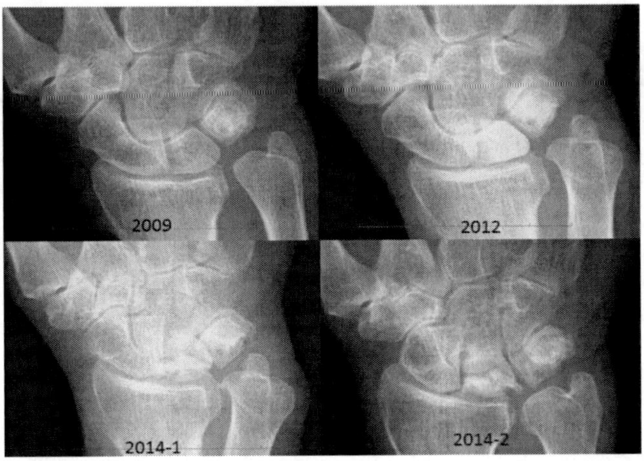

Figure 10. The PA views of the hand show sclerosis of the lunate caused by osteonecrosis that followed a glucocorticoid therapy. Bone collapse due to osteonecrosis of the lunate is observed.

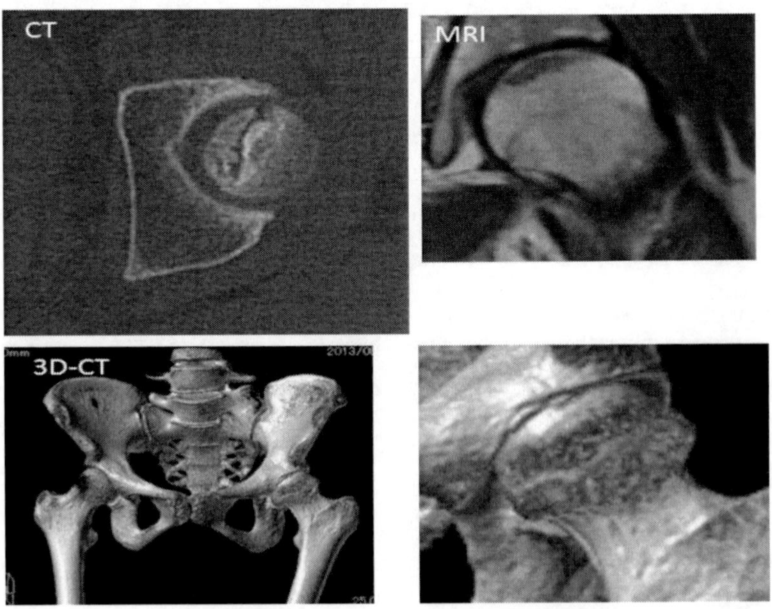

Figure 11. CT scan of the hip joint shows the structure of collapsed femoral head.

Bone Scintigraphy (Bone Scan)

Bone scintigraphy, technetium-labeled radionuclide bone scan, may be useful for the early diagnosis of osteonecrosis. Bone scintigraphy is a more sensitive test than plain radiography. However, bone scintigraphy is low resolution and non-specific. In the case of a contraindication to MRI, bone scintigraphy may be used.

CT: Computed Tomography

CT provides more detailed examination of the femoral head. CT has advantages to plain radiography in defining the extent of collapse in detail (Figure 11).

MRI: Magnetic Resonance Imaging

MRI can correctly reflect the pathological abnormalities of the femoral head [31, 32], and it is more sensitive than usual plain radiography in the diagnosis of osteonecrosis [17]. MRI detects silent osteonecrosis very early with the high sensitivity and specificity in most of the SLE patients treated with high doses of corticosteroids.

Early Changes in MRI

T1-Weighted Images

Osteonecrosis is a typical area of normal marrow signal bordered by a dark line or rim, low-intensity band (band-like pattern) on T1-weighted images (Figure 12 and 13). The band is attributed to the reaction at the interface between dead and viable bone.

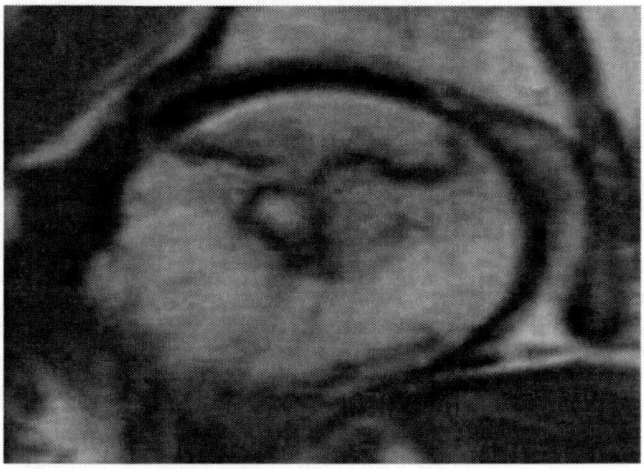

Figure 12. Band sign in T1-weighted MRI.

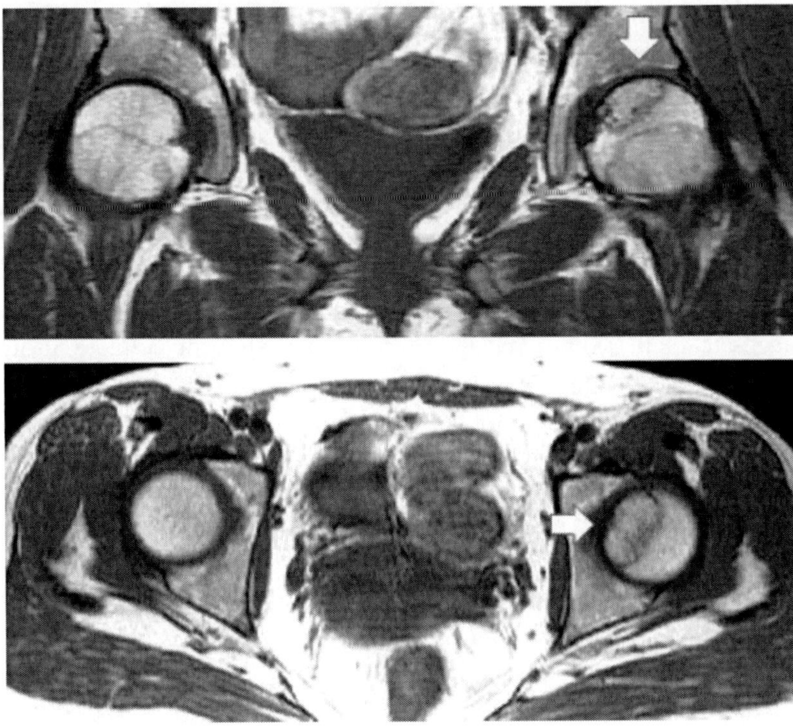

Figure 13. Band sign in T1-weighted MRI.

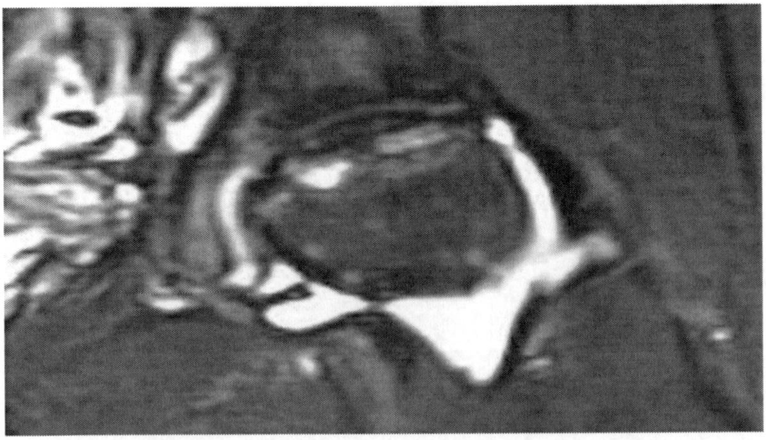

Figure 14. The double line sign in T2-weighted image.

T2-Weighted Images

On T2-weighted images, a narrower line of low signal intensity reflects the presence of bone sclerosis. The double line sign is seen in approximately 75% of cases [33]. An inner zone of high signal intensity indicates the location of granulation tissue (Figure 14).

Advanced Changes in MRI

In the advanced stage, subchondral fracture occurs just before collapse. The subchondral fracture line may be bright on T2-weighted and dark on T1-weighted MR images (Figure 15). The lesion of osteonecrosis may become dark on both T1- and T2-weighted MR images.

Bone Infarct in MRI

Bone infarct is dark on T1 and bright on T2-weighted sequences, with a geographic morphology. The T2 signal heterogeneity is not evident in bone infarcts.

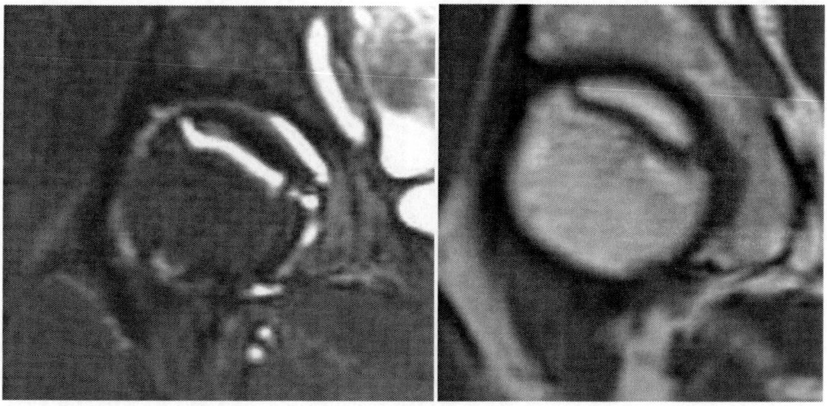

Figure 15. Subchondral fracture line in T2- and T1-weighted images.

G. Goal: Diagnosis

Although the diagnosis of osteonecrosis is made through clinical findings and various imaging modalities, the plain radiograph may be insensitive for earliest features of osteonecrosis. The clinical diagnosis of osteonecrosis may be difficult and be delayed. The diagnosis of osteonecrosis is more difficult in a patient with a prior osteoarthritis. Patients with SLE on high doses of corticosteroids heighten the probability of osteonecrosis.

The diagnosis of osteonecrosis should be always considered when an inactive patient with SLE treated with corticosteroids has persistent pain in the joint(s). Therefore, in this state, MRI is a utility technique for the diagnosis in various anatomic locations. The differential diagnosis includes spontaneous osteonecrosis of the knee, idiopathic transient osteoporosis of the hip, osteonecrosis of pregnancy/transient osteoporosis of the hip, transient regional osteoporosis, bone marrow edema syndrome (BMES), subchondral fracture, acute synovitis, soft tissue trauma, severe osteoarthritis, algodystrophy, and referred pain.

Preventive and Therapeutic Strategy for Osteonecrosis

There have been seldom studies of preventive strategy for osteonecrosis in SLE patients on high doses of corticosteroids. Although the general treatment of osteonecrosis in SLE patients is not so different from other causes, early diagnosis is important to the management of osteonecrosis.

Surgical Intervention

The conservative treatment includes analgesics and the use of devices to allow non-weight bearing. However, conservative approaches do not completely prevent progression. The patients with symptomatic lesions usually require surgical treatment.

There are various surgical interventions of femoral head including core decompression, structural bone grafting, vascularized fibula grafting, osteotomy, resurfacing arthroplasty, hemiarthroplasty, and total hip replacement (Figure 16 and 17). The timing and type of surgical interventions depend on the involved site and the stage of ONF.

Pharmacological Therapy

Drug Therapy

Because ONF is a serious complication in SLE patients associated with high doses of corticosteroids, the treatment and prevention of the disease are essentially required. Bisphosphonates, statins, anticoagulants, and vasodilators have been evaluated for the treatment of osteonecrosis of the femoral head. However, there have been few trials of prevention of osteonecrosis. Therefore, further studies are required to confirm their pharmacological effects.

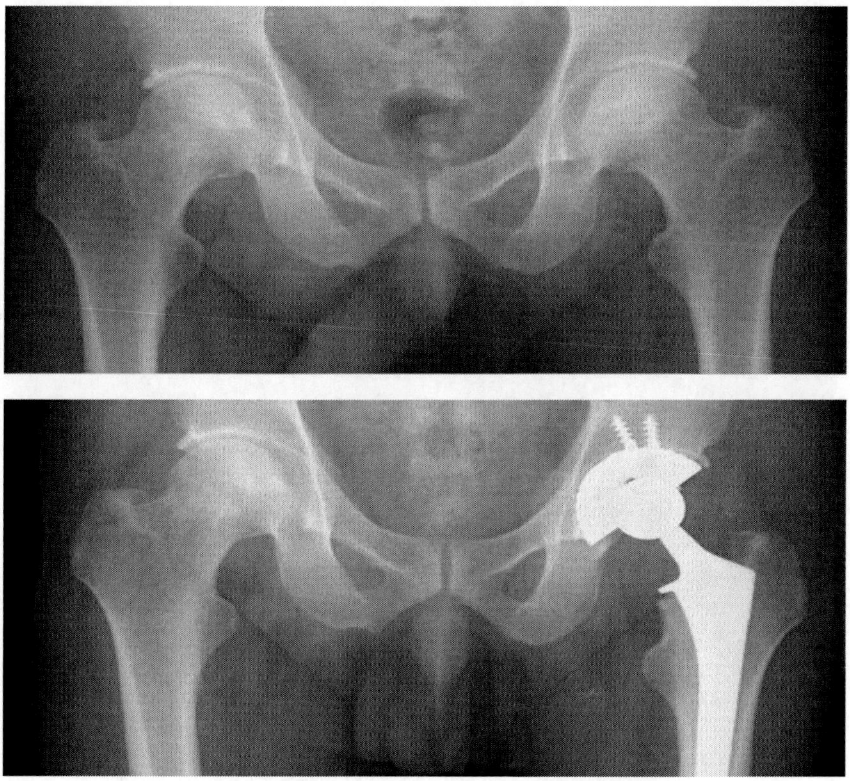

Figure 16. The osteonecrosis of the left femoral head had a significant impact on functional ability and required total joint replacement in a patient with SLE.]

Bisphosphonates

Bisphosphonates including alendronate may be beneficial in the treatment of osteonecrosis of the femoral head in some studies [34, 35]. An uncontrolled study of alendronates reported an improvement in the clinical function, a reduction in the rate of collapse, and a decrease in the requirement for total hip replacement [34]. A controlled but unblinded prospective study also showed a beneficial effect of alendronates in SLE patients on high doses of corticosteroids [35]. The proportion of collapsed hips in the alendronate treated group (6.9%; two of twenty-nine femoral heads) was significantly smaller than the control group (76%; nineteen of twenty-five femoral heads). As a result, joint replacement was necessary in only one patient of the alendronate treated group as compared to 16 cases of the control group. This result suggests that bisphosphonates appear to prevent the collapse of osteonecrosis in SLE patients. However, a randomized trial of oral alendronate did not prove a significant difference in the outcomes [36].

Bisphosphonate-induced osteonecrosis of the jaw is also important because of the frequent use of bisphosphonates on osteoporosis induced by corticosteroids.

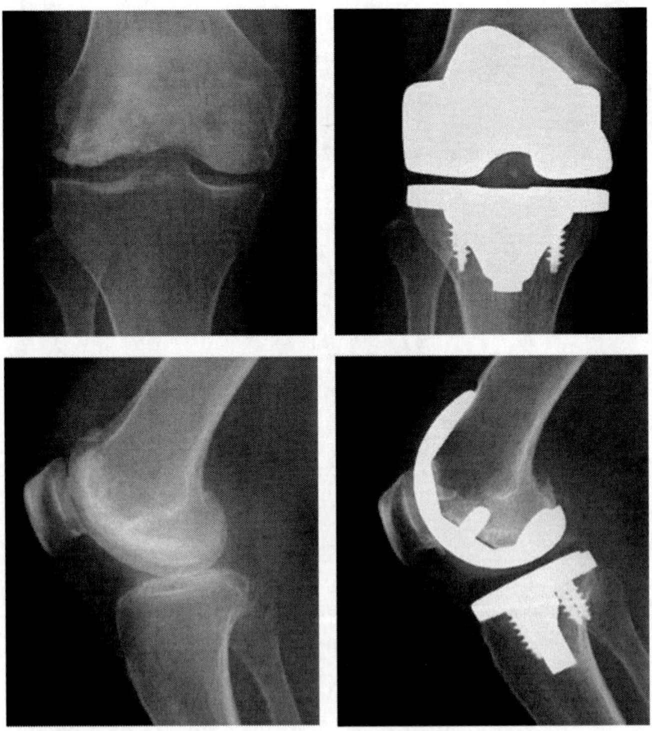

Figure 17. The osteonecrosis and total joint replacement of the right knee in a patient with SLE.

Anticoagulants

Anticoagulants including warfarin may prevent osteonecrosis. A multicenter prospective study of warfarin in 60 patients with SLE on high doses of corticosteroids was reported. The patients were assigned to either of two groups; a warfarin treated group and a control group [4]. In this study, warfarin has been continued at least for three months. Development of ONF in SLE patients has been observed by both plain radiography (symptomatic osteonecrosis) and MRI (silent osteonecrosis) for over five years.

Fewer hips in the warfarin treated group (21%) developed silent osteonecrosis compared to the control group (33%). Warfarin tended to prevent symptomatic osteonecrosis; 4.8 % in the warfarin treated group and 14% in the control group. However, the differences were not statistically significant.

Conclusion

In patients with autoimmune diseases treated with high doses of corticosteroids, early diagnosis of osteonecrosis using MRI and establishment of strategy of treatment are important issues. Further examination is required.

Abbreviation

AP view	Aanteroposterior view
ARCO	the Association of Research Circulation Osseous
CT	Computed tomography
AVN	avascular necrosis
MRI	magnetic resonance imaging
ONF	Osteonecrosis of femoral head
SLE	Systemic lupus erythematosus

References

[1] Chang, C; Greenspan, A; Gershwin, M. Osteonecrosis: current perspectives on pathogenesis and treatment. *Semin Arthritis Rheum*, 1993, 23, 47-69.

[2] Assouline-Dayan, Y; Chang, C; Greenspan, A; et al. Pathogenesis and natural history of osteonecrosis. *Semin Arthritis Rheum*, 2002, 32, 94-124.

[3] Koarada, S; Sadanaga, Y; Nagao, N; et al. Illustrated overview of the prevalence and clinical symptoms of systemic lupus erythematosus. *Systemic Lupus Erthematosus (SLE)*, 2013, Chapter1, 1-47. Nova Science publisher. New York.

[4] Nagasawa, K; Tada, Y; Koarada, S; et al. Prevention of steroidinduced osteonecrosis of femoral head in systemic lupus erythematosus by anti-coagulant. *Lupus*, 2006, 15, 354-357.

[5] Pistiner, M; Wallace, DJ; Nessim, S; et al. Lupus erythematosus in the 1980s: a survey of 570 patients. *Semin Arthritis Rheum.*, 21, 55-64, 1991.

[6] Abeles, M; Urman, J; Weinstein, A; et al. Systemic lupus erythematosus in the younger patient: survival studies. *J Rheumatol*, 1980, 7, 515-522.

[7] Diaz-Jouanen, E; Abud-Mendoza, C; Inglesias-Gamarra, A; et al. Ischemic necrosis of bone in systemic lupus erythematosus. *Orthop Rev*, 1985, 14, 303-309.

[8] Gladman, DD; Charudhry-Ahluwalia, V; Ibanez, D; et al. Outcomes of symptomatic osteonecrosis in 95 patients with systemic lupus erythematosus. *J Rheumatol*, 2001, 28, 2226-2229.

[9] Rascu, A; Manger, K; Kraetsch, H; et al. Osteonecrosis in systemic lupus erythematosus, steroid-induced or a lupus-dependent manifestation? *Lupus*, 1996, 5, 323-327.

[10] Tektonidou, M; Malagari, K; Vlachoyiannopoulos, P; et al. Asymptomatic avascular necrosis in patients with primary antiphospholipid syndrome in the absence of corticosteroid use: a prospective study by magnetic resonance imaging. *Arthritis Rheum*, 2003, 48, 732-736.

[11] Zizic, TM; Marcoux, C; Hungerford, DS; et al. The early diagnosis of ischemic necrosis of bone. *Arthritis Rheum*, 1986, 29, 1177-1186.

[12] Cozen, L; Wallace, DJ. Avascular necrosis in systemic lupus erythematosus: clinical associations and a 47-year perspective. *Am J Orthop (Belle Mead NJ)*, 1998, 27, 352-354.

[13] Oinuma, K; Harada, Y; Nawata, Y; et al. Osteonecrosis in patients with systemic lupus erythematosus develops very early after starting high dose corticosteroid treatment. *Ann Rheum Dis*, 2001, 60, 1145-1148.

[14] Weiner, ES; Abeles, M. Aseptic necrosis and glucocorticosteroids in systemic lupus erythematosus: a reevaluation. *J Rheumatol*, 1989, 16, 604-608.

[15] Massardo, L; Jacobelli, S; Leissner, M; et al. High-dose intravenous methylprednisolone therapy associated with osteonecrosis in patients with systemic lupus erythematosus. *Lupus*, 1992, 1, 401-405.

[16] Koo, K; Kim, R; Kim,Y; et al. Risk period for developing osteonecrosis of the femoral head in patients on steroid treatment. *Clin Rheumatol*, 2002, 21, 299-303.

[17] Nagasawa, K; Tsukamoto, H; Tada, Y; et al. Imaging study on the mode of development and changes in avascular necrosis of the femoral head in systemic lupus erythematosus: long-term observations. *Br J Rheumatol*, 1994, 33, 343-347.

[18] Nakamura, Y; Harada, K; Oinuma, S; et al. Spontaneous repair of asymptomatic osteonecrosis associated with corticosteroid therapy in systemic lupus erythematosus: 10-year minimum follow-up with MRI. *Lupus*, 2010, 19, 1307–1314.

[19] Ficat, RP; Arlet J: Ischemia and necroses of bone. 171-182, 1980 Williams & Wilkins Baltimore

[20] Steinberg, M; Hayken, G; Steinberg, D. A new method for evaluation and staging of avascular necrosis of the femoral head. pp. 398-403. *Arlet J Ficat R Hungerford D*. Bone circulation. 1984 Williams & Wilkins Baltimore.

[21] Gardeniers, JM. ARCO international classification of osteonecrosis. *ARCO Newsletter*, 1993, 5, 79–82.

[22] Hasan, SS; Romeo, AA. Nontraumatic osteonecrosis of the humeral head. *J Shoulder Elbow Surg*, 2002, 11, 281-298.

[23] Mitchell, DG; Rao, V; Dalinka, M; et al. MRI of joint fluid in the normal and ischemic hip. *AJR Am J Roentgenol*, 1986, 146, 1215-1218.

[24] Klippel, J; Gerber, L; Pollak, L; et al. Avascular necrosis in systemic lupus erythematosus: silent symmetric osteonecroses. *Am J Med*, 1979, 67, 83-87.

[25] Klipper, A; Stevens, M; Zizic, T; et al. Ischemic necrosis of bone in systemic lupus erythematosus. *Medicine*, 1976, 55, 251-257.

[26] Fishel, B; Caspi, D; Eventov, I; et al. Multiple osteonecrotic lesions in systemic lupus erythematosus. *J Rheumatol*, 1987, 14, 601-604.

[27] Guillaume, M; Brandelet, B; Peretz, A. Unusual high frequency of multifocal lesions of osteonecrosis in a young patient with systemic lupus erythematosus. *Br J Rheumatol*, 1998, 37, 1248-1249.

[28] Abeles, M; Urman, JD; Rothfield, NF. Aseptic necrosis of bone in systemic lupus erythematosus. Relationship to corticosteroid therapy. *Arch Intern Med*, 1978, 138, 750 754.

[29] Calvo-Alén, J1; McGwin, G; Toloza, S; et al. Ann Rheum Dis. Systemic lupus erythematosus in a multiethnic US cohort (LUMINA): XXIV. Cytotoxic treatment is an additional risk factor for the development of symptomatic osteonecrosis in lupus patients: results of a nested matched case-control study., 2006, 65, 785-90.

[30] Nagasawa, K; Tada, Y; Koarada, S; et al. Very early development of steroid-associated osteonecrosis of femoral head in systemic lupus erythematosus: prospective study by MRI. *Lupus*, 2005, 14, 385-90.

[31] Kalunian, KC; Hahn, BH; Bassett, L. Magnetic resonance imaging identifies early femoral head ischemic necrosis in patients receiving systemic glucocorticoid therapy. *J Rheumatol*, 1989, 16, 959-63.

[32] Hauzeur, JP; Sintzoff S, Jr; Appelboom, T; et al. Relationship between magnetic resonance imaging and histologic findings by bone biopsy in nontraumatic osteonecrosis of the femoral head. *J Rheumatol*, 1992, 19: 385-392.

[33] Kerr, R. Imaging of musculoskeletal complications of hemophilia. *Semin Musculoskelet Radiol*, 2003, 7, 127-136.

[34] Agarwala, S; Shah, S; Joshi, VR. The use of alendronate in the treatment of avascular necrosis of the femoral head: follow-up to eight years. *J Bone Joint Surg Br*, 2009, 91, 1013-1018.

[35] Lai, KA; Shen, WJ; Yang, CY; et al. The use of alendronate to prevent early collapse of the femoral head in patients with nontraumatic osteonecrosis. A randomized clinical study. *J Bone Joint Surg Am*, 2005, 87, 2155-2259.

[36] Chen, CH; Chang, JK; Lai, KA; et al. Alendronate in the prevention of collapse of the femoral head in nontraumatic osteonecrosis: a two-year multicenter, prospective, randomized, double-blind, placebo-controlled study. *Arthritis Rheum*, 2012, 64, 1572-1578.

In: Corticosteroids and Steroid Therapy
Editor: Carmen Adkins
ISBN: 978-1-63482-308-1
© 2015 Nova Science Publishers, Inc.

Chapter 5

The Correlation of Soluble Endothelial Protein C Receptor and High Dose Corticosteroid Therapy in Patients with Systemic Autoimmune Diseases

Syuichi Koarada*[*] *and Yoshifumi Tada
Division of Rheumatology, Faculty of Medicine,
Saga University, Saga, Japan

Abstract

Corticosteroids may sometimes exert unfavorable effects, thrombosis, avascular necrosis, endothelial cell damage, and corticosteroid vasculitis on the blood vessels in systemic autoimmune diseases. The association of corticosteroid and the thrombotic events has been reported. However, the mechanism of thrombotic tendency induced by corticosteroids has not been fully elucidated. Soluble endothelial cell protein C receptor (sEPCR) is one of the factors to regulate coagulation system. We found that sEPCR is a sensitive biomarker of endothelial injuries caused by active disease and often by corticosteroids in systemic

[*] Corresponding author: Division of Rheumatology, Faculty of Medicine, Saga University, 5-1-1 Nabeshima, Saga 849-8501, Japan; Tel: +81-(952)34-2367; Fax: +81-(952)34-2017.

lupus erythematosus (SLE). The findings of frequent events especially in the patients with SLE may illustrate the relationship between sEPCR and corticosteroids. sEPCR could be used as a predictable marker of vascular complications and thrombosis induced by corticosteroids in SLE.

Introduction

Systemic lupus erythematosus (SLE) is a systemic autoimmune disease with multiple organ involvements characterized by an inflammatory process and vascular disorders with abnormal immunological findings [1]. Cardiovascular and cerebrovascular diseases are much more common in patients with SLE than normal subjects [2, 3]. Particularly, vascular diseases in the patients younger than 40 years old are much common.

Protein C is an important vitamin K-dependent anticoagulant in the coagulant system. Protein C is synthesized largely in the liver [4]. The pathway of protein C is initiated by the binding of thrombin to thrombomodulin (TM). Then, the thrombin–TM complex converts protein C to activated protein C (APC) [5].

The endothelial protein C receptor (EPCR) is a type I transmembrane glycoprotein expressed in the endothelium and another member of the protein C anticoagulant pathway that enhances the activation of protein C [5, 6]. Although EPCR is expressed on the surface of endothelial cells (membrane bound EPCR; mEPCR) [7], mEPCR is mainly expressed on the surfaces of large vessels, especially arteries [8, 9].

The extracellular domain of EPCR is cleaved with a metalloprotease and soluble EPCR (sEPCR) is produced by the shedding of mEPCR [10] in endothelial injuries [11, 12]. Therefore, serum sEPCR levels reflect endothelial injuries. sEPCR may be a predictor of macrovascular complications and thrombosis in type 1 diabetes mellitus (DM) [13]. Because sEPCR and mEPCR bind to protein C and APC with the same affinity [14], sEPCR inhibits function of APC including the anti-inflammatory, profibrinolytic, and anticoagulant effects. Overproduction of sEPCR is a risk factor for thrombosis [15].

Endothelial injuries induce the atherosclerotic diseases in SLE and they develop with various risk factors, including longer duration of disease, vascular inflammation, hyperlipidemia, hypertension, and corticosteroids [16-18].

Systemic rheumatic diseases including SLE and vasculitis syndromes have endothelial injuries due to vasculitis. Moreover, corticosteroids may injure the endothelials. We review the reports of serum levels of sEPCR in patients with rheumatic disease and discuss the effect of corticosteroids to endothelial damage and possible association with thrombosis.

sEPCR Levels in Rheumatic Diseases

We performed the cross-sectional study of sEPCR levels in patients with SLE, various rheumatic diseases, and normal subjects [19]. The levels of sEPCR are higher in SLE patients than normal subjects [19]. The levels of sEPCR of patients with active SLE were higher than inactive patients. In another study, sEPCR levels were higher in SLE patients with renal involvement than patients without renal disease [20]. sEPCR levels correlate with serum creatinine. These results suggest that sEPCR levels are associated with disease manifestations and severity in SLE. There is a correlation between sEPCR levels and vascular dysfunction in SLE. It has been reported that the level of sTM is a marker reflecting endothelial cell injury [21], and elevated sTM levels are found in SLE patients with renal disorders.

Although the levels of both sEPCR and sTM are higher in SLE patients with active disease than inactive patients, sEPCR is more sensitive than sTM. sEPCR is a more sensitive biomarker of disease activity and vascular injuries in SLE.

In other rheumatic diseases including rheumatoid arthritis (RA), polymyositis and dermatomyositis (PM / DM), and adult onset Still's disease (AOSD), sEPCR levels are also higher than normal subjects [19]. These results suggest that endothelial injuries may also develop to some extent in these rheumatic diseases.

Effect of Corticosteroid on sEPCR in SLE

Corticosteroid treatment reduces the mean levels of sEPCR. Although 47% of SLE patients maintained high levels of sEPCR after corticosteroid therapy, in patients with other rheumatic diseases, elevated sEPCR levels were found only in 14% of patients after corticosteroid treatment [19]. The levels of sEPCR were raised even more by corticosteroids in some cases.

However, the mechanism of elevation of sEPCR levels in the patients with SLE after corticosteroid treatment is unrevealed. Although high doses of corticosteroids are effective in the treatment of SLE, they can also promote the endothelial damage in already existing vasculitis of SLE. Endothelial injuries induced by corticosteroids may lead to the development of atherosclerotic diseases in concert with vasculitis of SLE.

Thrombotic events occur in relation to active SLE and high doses of corticosteroids [22].

Also, among several factors, corticosteroid use is considered to be a major risk factor for development of avascular osteonecrosis in SLE patients [23-27]. Inhibition of the protein C and APC system by increased sEPCR may be related to thrombotic tendency.

Genetic Factors of sEPCR in SLE

Another mechanism for increased shedding of EPCR is a genetic factor, which may be associated with increased levels of sEPCR in SLE patients. The A/G genotype at exon 4 of the EPCR gene (A6936G) leads to elevated sEPCR in humans [28]. The high-shedding G/G genotype of EPCR is more prevalent in SLE patients than normal subjects [20]. Gene polymorphisms of EPCR may predict and reflect vascular injury in SLE. However, the relationship between gene polymorphisms and corticosteroids has not been clarified yet.

Conclusion

sEPCR levels were associated with disease manifestations and severity in SLE patients.

However, importantly, in some cases of SLE patients, sEPCR levels may remain elevated or even be increased by corticosteroids. Corticosteroids can promote the endothelial damage in already existing vasculitis in SLE. Further studies are required to understand the mechanism of the correlation of sEPCR and corticosteroids in SLE.

Abbreviations

APC	activated protein C
DM	diabetes mellitus
EPCR	endothelial protein C receptor
mEPCR	membrane bound EPCR
sEPCR	soluble endothelial cell protein C receptor
SLE	systemic lupus erythematosus
TM	thrombomodulin

References

[1] Koarada S. Autoantibody-producing B cells and B cell therapy in systemic lupus erythematosus. 55-83. In T. D. Marquez, D. U. Neto, editor: Lupus: Symptoms, Treatment and Potential Complications (Series: Immunology and Immune System Disorders), Nova Science Publishers, New York, USA, 2012.

[2] Morelli, S; Bernardo, M; Viganego, F; et al. Left-sided heart valve abnormalities and risk of ischemic cerebrovascular accidents in patients with systemic lupus erythematosus. *Lupus*, 2003, 1 2, 805–812.

[3] Bessant, R; Hingorani, A; Patel, L; et al. Risk of coronary heart disease and stroke in a large British cohort of patients with systemic lupus erythematosus. *Rheumatology*, 2004, 43, 924–929.

[4] Liaw, PC; Neuenchwander, PF; Smirnov, MD; et al. Mechanisms by which soluble endothelial cell protein C receptor modulates protein C and activated protein C function. *J Biol Chem*, 2000, 275, 5447–5452.

[5] Fukodome, K; Esmon, CT. Identification, cloning and regulation of a novel endothelial cell protein C/activated protein C receptor. *J Biol Chem*, 1994, 269, 26486–26491.

[6] Regan, LM; Stearns-Kurosawa, DJ; Kurosawa, S; et al. The endothelial cell protein C receptor: inhibition of activated protein C anticoagulant function without modulation of reaction with proteinase inhibitors. *J Biol Chem*, 1996, 271, 1 7499–7503.

[7] Stearns-Kurosawa, DJ; Kurosawa, S; Mollica, JS; et al. The endothelial cell protein C receptor augments protein C activation by the thrombin–thrombomodulin complex. *Proc Natl Acad Sci USA*, 1996, 93, 10212–10216.

[8] Laszik, Z; Mitro, A Taylor, FB Jr; et al. Human protein C receptor is present primarily on endothelium of large blood vessels: implications for the control of the protein C pathway. *Circulation*, 1997, 96, 3633–3640.

[9] Isermann, B; Sood, R; Pawlinski, R; et al. The thrombomodulin–protein C system is essential for the maintenance of pregnancy. *Nat Med*, 2003, 9, 331–337.

[10] Xu, J; Qu, D; Esmon, NL; et al. Metalloproteolytic release of endothelial cell protein C receptor. *J Biol Chem*, 1999, 275, 6038–6044.

[11] Kurosawa, S; Stearns-Kurosawa, DJ; Hidari, N; et al. Identification of functional endothelial protein C receptor in human plasma. *J Clin Invest*, 1997, 100, 411–418.

[12] Stearns-Kurosawa, DJ; Swindle, K; D'Angelo, A; et al. Plasma levels of endothelial protein C receptor respond to anticoagulant treatment. *Blood*, 2002, 99, 526–530.

[13] Zaghloul, A; Al-Bukhari, TA; Al-Pakistani, HA; et al. Soluble endothelial protein C receptor and high sensitivity C reactive protein levels as markers of endothelial dysfunction in patients with type 1 and type 2 diabetes mellitus: Their role in the prediction of vascular complications. *Diabetes Res Clin Pract*, 2014, 106, 597-604.

[14] Esmon, CT. The endothelial protein C receptor. *Curr Opin Hematol.*, 2006, 13, 382–385.

[15] Van Marion, V; Rosendaal, FR; Vos, HL; et al. Haplotypes of the EPCR gene, plasma sEPCR levels and the risk of deep venous thrombosis. *J Thromb Haemost*, 2004, 2,1 305–1310.

[16] Manzi, S; Meilahn, EN; Rairie, JE; et al. Age-specific incidence rates of myocardial infarction and angina in women with systemic lupus erythematosus: comparison with the Framingham study. *Am J Epidemiol*, 1997, 145, 408–415.

[17] Roman, MJ; Shanker, BA; Davis, A; et al. Prevalence and correlates of accelerated atherosclerosis in systemic lupus erythematosus. *N Engl J Med*, 2003, 349, 2399–2406.

[18] Nacach, AZ; Barr, SG; Magder, LS; et al. Damage in systemic lupus erythematosus and its association with corticosteroids. *Arthritis Rheum*, 2000, 43, 1801–1808.

[19] Koarada, S; Tsuneyoshi, N; Haruta, Y; et al. Effect of disease activity and corticosteroids on serum levels of soluble endothelial cell protein C receptor in patients with systemic lupus erythematosus. *Mod Rheumatol*, 2009, 19, 173-179.

[20] Sesin, CA; Yin, X; Esmon, CT; et al. Shedding of endothelial protein C receptor contributes to vasculopathy and renal injury in lupus: in vivo and in vitro evidence. Kidney Int, 2005, 68, 407–408.

[21] Boffa, MC; Karmochkine, M. Thrombomodulin: an overview and potential implications in vascular disorders. *Lupus*, 1998, 7, 120–125.

[22] Calvo-Alen, J; Toloza, SM; Fernandez, M; et al. Systemic lupus erythematosus in a multiethnic US cohort (LUMINA). XXV. Smoking, older age, disease activity, lupus anticoagulant, and glucocorticoid dose as risk factors for the occurrence of venous thrombosis in lupus patients. *Arthritis Rheum*, 2005, 52, 2060–2068.

[23] Cozen, L; Wallace, DJ; Avascular necrosis in systemic lupus erythematosus: clinical associations and a 47-year perspective. *Am J Orthop* (Belle Mead NJ), 1998, 27, 352-354.

[24] Cooper, C; Steinbuch, M; Stevenson, R; et al. The epidemiology of osteonecrosis: findings from the GPRD and THIN databases in the UK. *Osteoporos Int*, 2010, 21, 569-577.

[25] Mont, M; Glueck, C; Pacheco, I; et al.: Risk factors for osteonecrosis in systemic lupus erythematosus. *J Rheumatol*, 1997, 24, 654-662.

[26] Zizic, T; Marcoux, C; Hungerford, D; et al. Corticosteroid therapy associated with ischemic necrosis of bone in systemic lupus erythematosus. *Am J Med*, 1985, 79, 596-604.

[27] Mok, C; Lau, C; Wong, R. Risk factors for avascular bone necrosis in systemic lupus erythematosus. *Br J Rheumatol*, 1998, 37, 895-900.

[28] Saposnik, B; Reny, JL; Gaussem, P; et al. A haplotype of the EPCR gene is associated with increased plasma levels of sEPCR and is a candidate risk factor for thrombosis. *Blood*, 2004, 103, 1311–8.

Index

A

D

E

F

G

H

I

J

L

M

N

O

P

Q

R

S

T

U

V

W